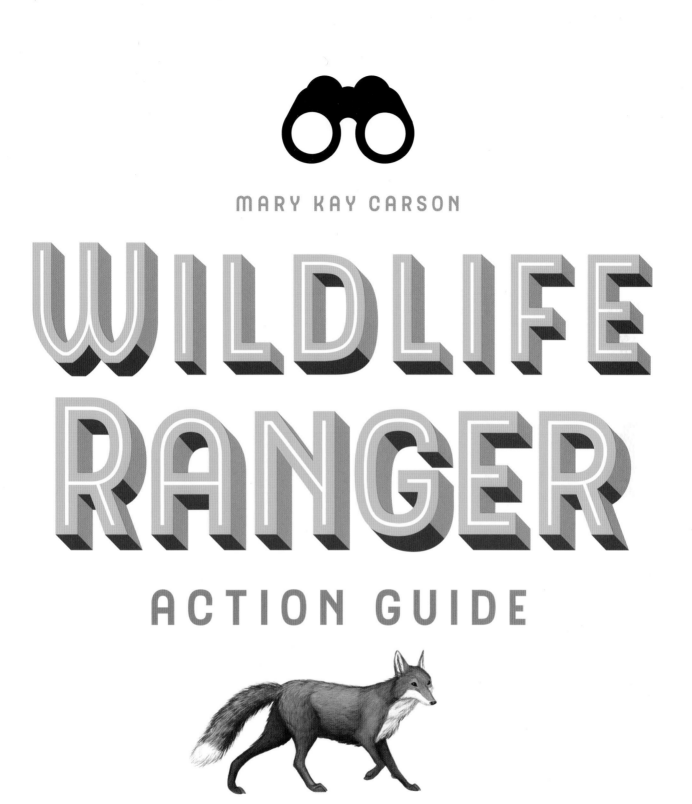

MARY KAY CARSON

WILDLIFE RANGER

ACTION GUIDE

Storey Publishing

The mission of Storey Publishing is to serve our customers by
publishing practical information that encourages
personal independence in harmony with the environment.

Edited by Deanna F. Cook and Lisa H. Hiley
Art direction by Jessica Armstrong
Book design and production by Stacy Wakefield Forte
Indexed by Nancy D. Wood

Cover photography by © Antagain/Getty Images, front (dragonfly); © Butterfly Hunter/Shutterstock, back (butterfly); © GlobalP/iStock.com, front (snake); © Jane Burton/Getty Images, front (squirrel); © joecicak/iStock.com, front (tortoise); © Ocs_12/ iStock.com, back (bird); © PeopleImages/iStock .com, front (child); © Ocs_12/iStock.com, back (bird); © sykadelx/Getty Images, back (deer); © Steve Maslowski, back (middle left); © SteveByland/iStock .com, front (bird); © sykadelx/Getty Images, back (deer); © Tom Uhlman, Tomuhlmanphoto.com, back (middle right)
Interior photography by © Tom Uhlman, Tomuhlmanphoto.com
Additional photography on inside back cover
Illustrations by © Jada Fitch
Range maps and tracks by Ilona Sherratt

Text © 2020 by Mary Kay Carson

Storey books are available at special discounts when purchased in bulk for premiums and sales promotions as well as for fund-raising or educational use. Special editions or book excerpts can also be created to specification. For details, please call 800-827-8673, or send an email to sales@storey.com.

Storey Publishing
210 MASS MoCA Way
North Adams, MA 01247
storey.com

Printed in China by Toppan Leefung Printing Ltd.
10   9   8   7   6   5   4   3   2   1

Library of Congress Cataloging-in-Publication Data

Names: Carson, Mary Kay, author.
Title: Wildlife ranger action guide / Mary Kay Carson.
Description: North Adams, MA : Storey Publishing, 2020. | Includes index. | Audience: Ages 6-12 | Audience: Grades 2-3 | Summary: "Hands-on activities and projects encourage children to learn about and take an active role in protecting local wildlife. Field guides covering 78 North American wildlife species teach kids about each and include tips for providing the plants and food needed for their survival"—Provided by publisher.
Identifiers: LCCN 2019033541 (print) | LCCN 2019033542 (ebook) | ISBN 9781635861068 (paperback) | ISBN 9781635861075 (hardcover) | ISBN 9781635861082 (ebook)
Subjects: LCSH: Animals—United States—Identification—Juvenile literature.
Classification: LCC QL49 .C337 2020 (print) | LCC QL49 (ebook) | DDC 591—dc23
LC record available at https://lccn.loc.gov/2019033541
LC ebook record available at https://lccn.loc .gov/2019033542

This book is dedicated to the wild creatures of Chuckland, the backyard nature sanctuary of the author and photographer Tom Uhlman.

A BIG THANKS
to our terrific kid helpers and wildlife photographer Steve Maslowski.

4

5

6

·1·
YOU CAN SAVE
WILD
ANIMALS!

GRASSHOPPER

RED SQUIRREL

WILDLIFE NEEDS YOUR HELP! You may think of wildlife as being African elephants and Antarctic penguins, but there's wildlife all around you. How can you save the wild animals, like birds, foxes, butterflies, and turtles, that live in your neighborhood?

By becoming a backyard ranger and making your own yard more wildlife friendly! You can do this by creating habitats — places where butterflies find food, frogs seek shelter, nuthatches nest, or deer drink water. Caring for close-by creatures is exciting and fun, plus it's super important. Here are some reasons why:

- **LACK OF SPACE.** Just like us, wild animals need places to live. But when people pave over prairies and cut down forests, the open space and habitat that wildlife needs are lost. When chunks of natural land are separated by highways, streets, buildings, and towns, the result is habitat fragmentation.

GREAT HORNED OWL

MAKE YOUR
YARD A BETTER
PLACE FOR
WILD ANIMALS
AND WATCH AS
WOODPECKERS,
DRAGONFLIES,
CHIPMUNKS,
TOADS, OR
OTHER ANIMALS
SHOW UP.

- **BROKEN-UP HABITATS** make it harder for animals to find enough food, water, and shelter to survive. A large, connected habitat supports more species (types) of wildlife than a bunch of divided ones. More species of animals and plants equals biodiversity, which is better for the environment.

- **LACK OF FOOD.** Most lawns, farms, and city parks aren't natural habitats. The grass, vegetables, trees, and bushes that grow there are often non-native plants imported from other continents, countries, or regions. Some are harmful invasive plants. Animals depend on specific native plants for shelter, nesting places, and food.

For example, monarch caterpillars only eat milkweed. If non-native plants crowd out all the milkweed, the caterpillars starve and don't become butterflies. Then the birds that eat the butterflies go hungry, too. Non-native plants mess up the food web for everyone!

GREEN FROG

WHO EATS WHAT?

A fox eats a bird that ate a snail that ate a plant.

This food chain shows how energy passes from plants up through animals.

An ecosystem is a combination of many crisscrossing food chains called a food web.

HUMAN-CAUSED HAZARDS. Many farmers, homeowners, and gardeners use chemicals that harm insects and pollute water. Glass buildings and windows confuse birds, who crash into them. Cars and trucks run over animals on highways. Pet dogs and cats kill wildlife, too. Rats, starlings, and other non-native animals also harm wildlife by taking over needed habitat.

CENTIPEDE

Creating Natural Habitats = Saving Wild Animals

Now you know why wildlife needs your help. It's time to find out what you can do. Here's the secret formula, the Big Four ingredients of wildlife habitats: food, water, shelter, and nests.

1. **FOOD.** Do the right kinds of seeds or fruits grow nearby? Is there enough prey for animals to hunt?

2. **WATER.** Can animals find safe drinking water year-round?

3. **SHELTER.** Is there cover to hide from predators? Can animals find protection from rain and cold? Are there safe sleeping sites?

4. **NESTS.** Are there places for animals to raise their young? Are the right materials around to build nests?

One backyard can't provide all of the Big Four for every animal, but if we each do a little, it adds up to big differences. What you do in your backyard to help wildlife helps animals in the whole neighborhood. You can find lots of fun ways to share your yard with fawns, chickadees, butterflies, chipmunks, and ladybugs. Let's get started!

PORCUPINE

MALLARD DUCK

1. FOOD

2. WATER

3. SHELTER

4. NESTS

WHO LIVES NEAR ME?

Are there toads hiding in your garden? Shrews scurrying under backyard leaves? A couple of crows hanging out in a high branch of a nearby tree? The first step to helping local wildlife is to find out which animals live around you.

Remember, all animals need access to safe water, suitable food, and the right kind of shelter and nesting spaces. Look for wildlife where there's water, food, and shelter in your neighborhood.

BE WILDLIFE AWARE

Exploring your yard is exciting and enjoyable. Keep it that way by knowing what to avoid.

- Beware of itchy, rash-causing plants like poison ivy and poison oak. *Leaves of three, leave it be!*

- Watch out for bugs that can sting or bite like wasps, bees, ticks, and centipedes. If you're not sure, don't pick it up.

- Are there venomous snakes where you live? Learn what they look like.

- Hands off wild furry animals! They can carry rabies, a serious disease. If a raccoon or bat isn't fleeing when it sees you, it's likely sick or injured. Find a grown-up.

CROW

SHREW

USING THIS BOOK

Once you have an idea of the kinds of animals living around you, you can improve the habitat of the current creatures as well as plan for future visitors. This book is all about observing, improving, and creating natural habitats. Each chapter has information, activities, and projects designed to attract and support birds, bugs, reptiles, amphibians, and mammals.

Look for these icons on the pages for extra tips and ideas.

BECOME A CITIZEN SCIENTIST!

Do you want to learn more and do more to save wild animals? People of all ages are taking action to help wildlife everywhere. Citizen scientists can assist professional scientists in many ways. One example is helping researchers gather data by reporting how many frogs or birds are spotted in a particular area from year to year.

Check out page 175 for ways to join the greater conservation community. You're not alone!

Links to woodworking plans and printable pages that help you take it to the next level

Suggestions for what to note, list, sketch, or document in your Wild Notes notebook (page 15)

CHECK THIS OUT!

Information about making your backyard ecosystem more welcoming and planting native plants to attract wildlife

THE FIELD GUIDE PAGES at the ends of chapters 2 through 6 will help you identify which birds are feasting at the feeders, which insect just buzzed by your ear, and what fluffy thing disappeared down that hole. The pages are color-coded by chapter and look like this:

Scientific name to use when looking up further information at the library or online

Size, color, sounds, and other clues for quick identification

Map showing where the animal lives in North America

Places you're likely to see this animal

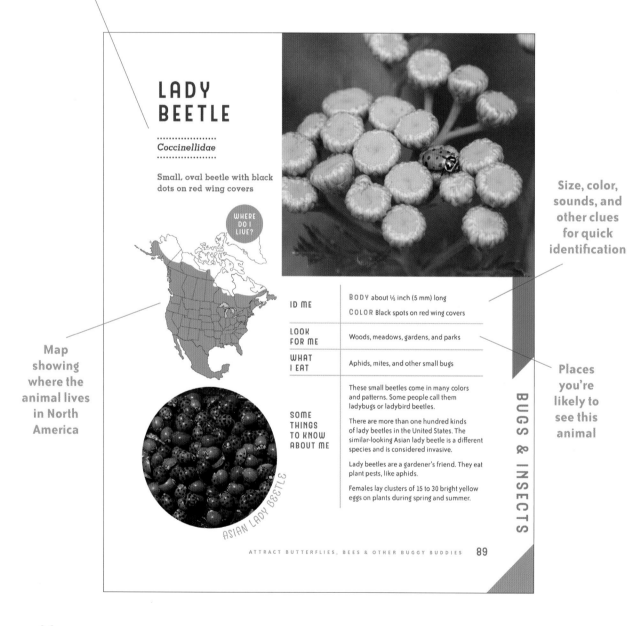

LADY BEETLE

Coccinellidae

Small, oval beetle with black dots on red wing covers

WHERE DO I LIVE?

ID ME	BODY about ⅕ inch (5 mm) long
	COLOR Black spots on red wing covers
LOOK FOR ME	Woods, meadows, gardens, and parks
WHAT I EAT	Aphids, mites, and other small bugs
SOME THINGS TO KNOW ABOUT ME	These small beetles come in many colors and patterns. Some people call them ladybugs or ladybird beetles.
	There are more than one hundred kinds of lady beetles in the United States. The similar-looking Asian lady beetle is a different species and is considered invasive.
	Lady beetles are a gardener's friend. They eat plant pests, like aphids.
	Females lay clusters of 15 to 30 bright yellow eggs on plants during spring and summer.

ASIAN LADY BEETLE

BUGS & INSECTS

ATTRACT BUTTERFLIES, BEES & OTHER BUGGY BUDDIES   89

Making the world — including your yard — a better place for animals takes time and patience. Plants don't grow overnight. Bats and birds need time to move in. Every backyard is different. Improving your particular habitat will take some experimenting. That's why it's important to keep track of your progress.

TAKING NOTES

Create a Wild Notes notebook for recording your observations, projects, and successes. Use any kind of notebook with plenty of blank pages. A ringed binder works well because you can add pages to it with a hole puncher. If you're a more techy kind of kid, you could create a Wild Notes file folder on the computer or use a journaling app.

 What should go in your notebook? Look for suggestions throughout the book. Here are some more ideas:

- Landscape-style map of your yard for planning

- Notes on successes, failures, and ideas for improvement

- Lists of animals seen with dates, time of day, and locations

- Sketches or photos of unknown birds and other animals to try to identify

- Notes on which kinds of feeders and foods seem to attract which animals

- Photographs or fact sheets from other sources

- Nature poetry and art

Print out some starter pages for your notebook. Look for the web address on page 175.

WILD NOTES NOTEBOOK

MATERIALS

2 or 3 outdoor chairs

A large tarp or several yards of mesh or netting

Clothespins or large binder clips

Scissors

Binoculars (optional)

WILDLIFE SPOTTING BLIND

Wild animals are shy of people. Building a blind, or a place where you can hide, will camouflage you, making the creatures in your yard easier to spy on. Happy wildlife spotting!

HOW TO

1 Set up some chairs and drape the tarp or netting over them. Use clothespins to clip the tarp ends onto the chairs. Make sure you leave an opening so you can get in and out!

2 Cut some peek-through flaps in the sides of the tarp. Check that they're at the right height from the inside.

3 Gather some leaves, twigs, and thin branches and use them to camouflage the fort.

4 Start watching! Snap photos and take notes on what you see. Make sure to include the time of day and the date.

WATER WATCH

Try attracting different animals by offering a source of water.

Gather a few planter saucers, old dinner plates, or other shallow dishes.

Place them around the yard at different heights — on the ground, on a table or stump, or wedged between bush branches. Fill the saucers with water and check water levels each day.

Put a couple near your Wildlife Spotting Blind, if you made one.

Rocks provide landing spots for butterflies and other smaller visitors.

TEST FEEDERS

Putting out test feeders is a great way to discover which birds are around and what their favorite foods are.

Place a few near your Wildlife Spotting Blind, if you made one.

Juice jugs, milk cartons, soda bottles, yogurt cups, and other recycling bin goodies make perfect temporary test feeders.

Wash all containers thoroughly with hot, soapy water and let dry completely before filling.

Hang some using yarn or string and set others on the ground or on a picnic table or large rock.

Use different types of seed in each feeder to see which attracts the most birds.

Add a dowel for a landing perch.

Cut holes in larger containers, leaving a lip to hold the seeds.

Keep notes on which birds are attracted to which feeders.

19

MATERIALS

Scissors

½-gallon (2L) coated milk or juice carton

Packing or duct tape

Felt

Waterproof glue

Peanut butter

Small paintbrush

Washable, nontoxic poster paint in a dark color

White paper

Paper clips

LITTLE CRITTERS PAW PRINTER

Mice, voles, shrews, and other small animals like to be hidden. They stay safe by traveling along pathways belowground or under tall grass. This critter tunnel captures tiny tracks so you can find out who's scurrying unseen around your yard.

HOW TO

1 Cut off the top and bottom of the carton. Cut off one of the sides; this becomes the printing tray. Fold the carton into a triangle and tape it closed to make the tunnel.

2 Trim the leftover carton side until it easily slides in and out of the tunnel. Cut two strips of felt, each 1 inch (2.5 cm) wide and as long as the printing tray is wide. Glue them near the middle of the tray, leaving an inch or so gap in the very center. That's where the bait goes! Let dry.

3 Smear a line of peanut butter between the felt strips. Use the paintbrush to soak the felt with a thick layer of paint.

4 Cut two rectangles of paper, one to fit on each end of the printing tray. Attach with paper clips. Carefully slide the tray into the tunnel.

Set your paw printer on the ground at dusk near some trees or alongside a patio or fence line. Check for tracks after a few hours or in the morning.

WHO TRAVELED THROUGH YOUR TUNNEL?

Mouse

Shrew

Chipmunk

Squirrel

SANDPIT TRACKS TRAP

Just because you don't see them doesn't mean they're not there. Animal tracks are proof positive that creatures are around. Another way to trap some tracks is to make a sandpit.

Before you go to bed, spread a thick layer of sand on the ground.

Set out a small dish of pet food.

Photograph or draw the tracks to help you identify who made them.

SKUNK

OPOSSUM

RACCOON

CAT

Check the sandpit for tracks the next morning.

FOX

DOG

Look overhead and up into trees for birds. Are there any nests around?

SAFARI SURVEY

Grab a hand lens, binoculars, and your notebook and head out on a backyard safari.

Look for leftovers like nutshells, clipped plants, slug slime trails, owl pellets, and scat, which is the scientific word for animal poop.

Turn over rocks and logs, look under fallen leaves, and study puddles to hunt for bugs, toads, and salamanders.

Scan for signs that animals are about. Check for footprints, feathers, bits of fur, scratches on tree trunks, and flattened patches of grass.

23

BENEFICIAL BUG BAIT

Gardeners like insects that eat plant pests such as aphids. To lure beneficial bugs, like this praying mantis, into their gardens, they put out attractants. That's just a fancy word for bug bait! Mix up your own and see what happens.

Thoroughly mix the following ingredients. Pour the mixture into a clean spray bottle.

- 5 teaspoons (25 mL) sugar
- 5 teaspoons (25 mL) powdered yeast
- 1 cup (240 mL) warm water

Spray the bug bait onto bushes and plants around the yard. Tie ribbons or yarn to mark where you're spraying.

Hang a white bedsheet from a strung rope, over a fence, or between some tree branches. Choose a dark area away from street or porch lights.

Lots of insects like the nightlife! Get a glance at who's flying around in your yard after dark by setting up a light trap.

NIGHTLIFE LIGHT TRAP

Watch the moths and beetles fly toward the light!

Shine a strong flashlight on the sheet. A portable ultraviolet lamp, also called a black light, will attract a wider variety of bugs.

Record the nocturnal visitors in your Wild Notes notebook and include photos if possible.

25

GULL

· 2 ·

COME ON IN,

FEATHERED FRIENDS

BIRDS ARE FUN to watch and identify. They're colorful, lively, and musical. Most birds are also out and about during the day, making them easier to observe than nocturnal (nighttime) animals. And they're right in your backyard.

SAVANNAH SPARROW

BIRDS NEED A LOT OF FOOD. They're warm-blooded and must eat to burn enough energy to keep their body temperature steady. Different birds eat different foods. Hummingbirds sip nectar, hawks hunt squirrels, wrens catch bugs, and finches snack on seeds. You need to know who eats what to create good bird habitat. A source of water will attract a variety of birds.

FOOD, WATER, AND SHELTER HELP BIRDS THAT ONLY STAY FOR A SEASON OR ARE JUST PASSING THROUGH, TOO.

BIRDS LAY EGGS, and most birds build a nest to lay them in. Good bird habitat has places to nest and provides the materials needed to make the nest — sticks, grasses, lichen, mud, and so on. Birds use bushes, trees, and other plants to shelter from bad weather and to hide from predators. These safe spots are called cover.

HOUSE WREN

27

BLUE JAY
FEATHERS

NATURAL HABITAT IS IMPORTANT for more than the birds that live there year-round. Many birds migrate, traveling hundreds of miles between summer and winter homes. Migrating birds can't get where they're going without safe stopovers. They need to fuel up on food and clean water.

BIRDS BATTLE MANY THREATS besides losing natural habitat. Cats and dogs kill them, cars and chemicals harm them, and they often crash into windows and glass-covered buildings. But you can help birds fight back! Each problem is an opportunity to help. Check out the rest of the chapter for some ways to save birds.

SPARE SOME HAIR?

Many birds collect animal hair to line their nests. The next time you brush Spot or Socks, toss the fur outside near some bushes or at the foot of a tree. It may become part of a cozy nest! Note: If your pet's fur has been treated with flea or tick pesticides or other chemicals, it's not safe for birds. Put it in the garbage. Human hair isn't great for nests either. It's so thin that it can hurt birds if it gets wrapped around their legs or necks.

HOUSE
SPARROW

BIO BASICS
PARTS OF A
BIRD

vane

shaft

eye ring

crown

nostril

beak

nape

throat

wing

breast

side

belly

tail

rump

feet

MATERIALS

Ruler

Large plastic coffee or other container with lid

Permanent marker or pen

Pointy-tipped scissors

Paint (optional)

Nail

String

COFFEE TUB NEST BOX

Many birds nest in holes, or cavities, in trees. Nest boxes are a great way to give cavity nesters more places to lay eggs and raise chicks. The trick is to cut the perfect-sized entry hole for the kind of bird you want to move in. So choose a bird, grab a plastic tub, and measure twice before cutting the hole.

HOW TO

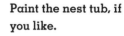

1 Choose a hole size from the chart. Measure and draw the same-sized hole between the center and the edge of the lid (not right in the middle).

2 With adult help, cut out the hole.

3 Paint the nest tub, if you like.

HOLE CHART

FOR THESE BIRDS...	...CUT THIS SIZE HOLE
House Wren, chickadees	1⅛ inches (2.9 cm)
Titmouse, some warblers, Red-breasted Nuthatch	1¼ inches (3.2 cm)
Swallows, White-breasted Nuthatch	1⅜ inches (3.5 cm)
Eastern and Western Bluebird, Carolina Wren	1½ inches (3.8 cm)
Flycatchers, Mountain Bluebird	1⁹⁄₁₆ inches (4 cm)

MEASUREMENTS ARE DIAMETERS.

Want to make a wooden nest box? Look for the web address on page 175.

4 Poke two holes in the side of the tub with a nail. Slip a length of string through the holes and tie it in a loop. Hang the nest box in a protected spot.

ANTI-CRASH WINDOW CLINGS

Every year millions of birds die when they accidently fly into windows, crashing into the glass. It's a big problem! One way to help is by making window glass easier for birds to see. These fun-to-make window clings can prevent bird crashes.

HOW TO

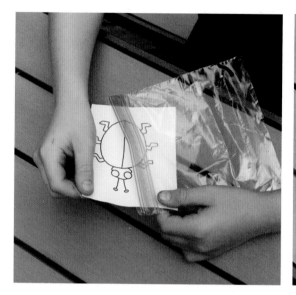

1 Draw or print a simple picture and slide it into a bag.

2 Apply the paint on the plastic bag, covering your picture. Make sure that all the lines are filled in and connected. Make a thick layer of paint.

For patterns, look for the web address on page 175.

3 Let the paint dry completely. Overnight is best.

4 Carefully peel off the clings and stick them to the inside of your window. The bigger the window, the more clings you can hang.

MATERIALS

Large nail

Disposable water bottle
with lid

Sharpened pencil

Plastic drinking straw
(red or other bright color)

Scissors

Outdoor glue or
waterproof caulk

String or yarn

Red decorations

½ cup (120 mL)
extrafine sugar

2 cups (480 mL)
hot water

HOMEMADE HUMMER FEEDER

Hummingbirds need lots of food to fuel their high-energy lives. Make this simple hummingbird feeder and some sugary nectar to help them tank up. Hang the feeder on a low branch or a plant hanger hook and see how many hummers show up.

HOW TO

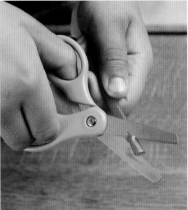

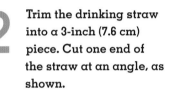

1 Use the nail to start a hole near the bottom of the water bottle. Make the hole bigger with a sharpened pencil. It should just be big enough to squeeze the drinking straw into it, but no wider.

2 Trim the drinking straw into a 3-inch (7.6 cm) piece. Cut one end of the straw at an angle, as shown.

3 Push the straight end of the straw into the hole. Glue around the hole to hold the straw in place with the angled end up. Let dry.

4 Tie a string around the neck of the bottle for hanging. Decorate your feeder with red stickers, fake flower petals, or markers. Hummingbirds love red!

5 Mix the sugar into the hot water until dissolved. (Don't add food coloring — it isn't good for the hummers.)
To fill the feeder, hold it in the sink with the straw pointing up and the neck of the bottle tilted. Pour in the nectar and screw on the lid before tipping the bottle upright — if you don't, all the nectar will flow out!

HUMMER SWING

Give fun-to-watch hummingbirds a place to perch between sweet sips.

Twist a pipe cleaner in the middle to make a loop.

Attach the end to an unsharpened pencil or short stick to make a swinging perch.

Hang the perch near a hummingbird feeder.

Tie the open end into a knot.

Grab an old sock and fill it with nyjer seeds.

Poke a few small holes in the sock.

SEEDY SOCK

Finches especially love tiny black nyjer seeds (wrongly called thistle seed sometimes). See if you can attract them with this easy-to-make feeder.

Hang it up and watch the finches fly in!

BIRD ALERT SCRUNCHY

MATERIALS

20 x 5-inch
(51 x 13 cm) piece
of lightweight fabric
with a colorful
pattern

Waterproof
fabric glue

Scissors

Reflective tape,
¼ to ½ inch wide
(0.6 to 1.3 cm)

Cat collar
(a breakaway
safety-buckle
version is best)

Cuddly cats are great pets. But to birds and other wildlife, your kitty is a predator armed with claws and teeth. This collar cover can help warn birds that danger is on the prowl. The bright colors and reflective strip catch the eye of birds, making it harder for cats to surprise them.

HOW TO

1 Lay the fabric on your work surface wrong side up. Fold down an inch (2.5 cm) or so of one long side of the fabric. Crease it with your fingers.

2 Fold the other long side so that it overlaps the first folded edge by about ½ inch (1.5 cm). Crease the edge. (You may need to iron it.)

buckle

3 Open the wider fold and run a line of glue along the edge of the first fold. Press the upper layer of fabric firmly onto the glue to form a tube. Let dry thoroughly.

4 Cut a 19-inch (48 cm) long strip of the reflective tape. Peel off the backing as you stick it on over the glued seam. If it isn't sticking well, use more glue.

5 Slip the collar through the fabric tube, scrunching up the tube and turning the tape to face the outside. It's ready for kitty to wear!

MATERIALS

Water

Microwaveable mixing
bowl and spoon

2 packets of
unflavored gelatin

2 tablespoons
(35 mL) corn syrup,
molasses, or honey

2½ cups (600 mL)
wild birdseed mix

Several large cookie
cutters or molds

Scissors

Plastic drinking straws
cut into short pieces

Yarn or string

ORNAMENTS FOR THE BIRDS

Nothing brightens up bare winter branches like a
collection of bird-food ornaments. You can smear
nut butter on pinecones or toilet paper rolls and
coat with wild birdseed or thread oat cereal loops
onto pipe cleaners. As an extraspecial treat for
your feathered friends, hang up a few of these
homemade bird cookies.

HOW TO

1 Pour ½ cup (120 mL) cold water into the bowl. Add both packets of gelatin and stir. Heat ½ cup water in the microwave until boiling. Add it to the mixture and stir until all the gelatin dissolves.

2 Add the sweetener and birdseed. Mix well. Set the bowl in the refrigerator to cool down for about 30 minutes.

3 When the mixture has firmed up, spoon it into the molds or cookie cutters. Press the mixture in firmly.

BIG BIRD FOOD

Are squirrels emptying your bird feeders? Don't stress about it too much. Squirrels are food for hawks and owls, so you're still feeding birds!

4 Poke a straw into each mold to create a hole near the center. Let the bird cookies set up overnight in the refrigerator.

5 Carefully pop the cookies out of the molds and remove the straws. Thread yarn or string through the holes and tie to make loops. Hang them outside and see who comes to visit.

QUICK- AND-EASY SUET CAKES

Winter birds flock to suet feeders. Suet is solid beef or mutton fat; it's a high-energy food that gives birds fuel to stay warm in cold weather. Fill suet feeders yourself with this recipe.

HOW TO

 Cut the lard into chunks, place into the bowl, and microwave on low until it starts to melt, 1 or 2 minutes. Add the peanut butter and microwave for another 1 to 2 minutes. Stir to combine.

 Mix in the oats, sunflower seeds, flour, cornmeal, and raisins.

3 Spread the gooey mess into a loaf pan. Place the pan in the freezer for 3 to 4 hours.

4 Once the suet is solid, cut it into feeder-sized squares.

5 Load a wire feeder or mesh bag with the suet and hang it up.

A DEAD TREE CAN BE FULL OF LIFE

A dead or dying tree may not look pretty, but it has a place in the ecosystem. If you have one in your yard that isn't a danger, ask your parents to leave it be! Lots of birds depend on dead trees. They're full of tasty insects for woodpeckers. And owls, wood ducks, bluebirds, wrens, and chickadees build cavity nests in dead trees.

WILD NOTES BIRDS

If you're wondering where to start with your notebook, here are some suggestions.

- In addition to making notes, make drawings or take photos of birds you see.

- Use the Field Guide pages to help you identify birds. Note how many you see of each species.

- Project updates are important! What's working, what's not?

- Which bird species are using which feeders? Is there a favorite food? Which are using the nest boxes?

- Start a birding "life list," a list of every species you've seen.

- Keep a bird calendar, writing down the dates when different migrating birds show up.

BIRD-FRIENDLY GARDEN

Native plants help your habitat by providing food, shelter, and nest sites for birds (and other animals). Trees like cedars, spruces, and oaks offer cover and nesting spaces. Shrubs like sumac, serviceberry, buttonbush, elderberry, and dogwoods as well as vines like Virginia creeper and wild grapes are food factories. Milkweeds, native sunflowers, coneflowers, and cardinal flowers are great for seeds and nectar. Use the native plant websites on pages 174–75 to help you choose what grows best where you live.

RUBY-THROATED HUMMINGBIRD

You can listen to bird calls on several different websites (see page 174). There are some links to birdsong apps there, too.

WHO'S THAT SINGING?

You often hear birds before you see them. Bird watchers learn to identify birds by both song and sight.

Use a recording device or phone app to record birdsong and calls in your yard.

AMERICAN CROW

Corvus brachyrhynchos

Large, all-black bird in treetops or flying in a noisy flock

WHERE DO I LIVE?

Breeding
Year-round
Winter

ID ME	SIZE 16 to 21 inches (40 to 53 cm) long
	COLOR Black body, eyes, bill, and legs
	VOICE Loud *caw-caw-caw-caw*
LOOK FOR ME	Parks, pastures, farmland, landfills, yards, and along rivers
WHAT I EAT	Seeds, nuts, worms, mice, insects, fish, bird eggs and nestlings, garbage, and carrion (dead animals)
SOME THINGS TO KNOW ABOUT ME	Crows are very social. Young birds stay with parents for years and help raise younger siblings.
	Crows often band together to run off predators, like hawks, a behavior called mobbing.
	In some places, flocks of crows gather to spend the night in huge roosts, sometimes numbering in the tens of thousands.

AMERICAN GOLDFINCH

Spinus tristis

Little, yellow-and-black bird with a short bill balancing on a seedpod

WHERE DO I LIVE?

- Breeding
- Year-round
- Winter

ID ME	**SIZE** 4½ to 5 inches (11 to 13 cm) long **COLOR** Males have bright yellow body in spring and summer. Females are always dull colored as are males during cold months. All have black forehead and wings. **VOICE** Squeaky *po-ta-to-chip* call while flying
LOOK FOR ME	Weedy fields, roadsides, backyards, farms, and seed feeders
WHAT I EAT	Seeds from flowers, grasses, trees, and feeders
PLANT THIS FOR ME	Native thistle, milkweed, asters, and sunflowers
SOME THINGS TO KNOW ABOUT ME	American Goldfinches breed later than most native birds. They don't start building nests until there are plenty of seeds to feed young, usually well into June or July. Goldfinches live in small flocks that fly in an up-and-down wave pattern. They're acrobatic eaters, snacking on seeds while hanging upside down on moving plant stalks. American Goldfinches molt their feathers twice a year, in late winter and late summer. Males switch to bright yellow for spring mating season.

BIRDS

AMERICAN ROBIN

Turdus migratorius

Large, dark songbird with orange chest searching through leaf litter or on lawns

WHERE DO I LIVE?

Breeding
Year-round
Winter

ID ME	SIZE 8 to 11 inches (20 to 28 cm) long
	COLOR Grayish brown wings and tail, darker head, orange chest, yellow beak
	VOICE Musical string of repeating whistles: *cheerily, cheer up, cheer up, cheerily, cheer up*
LOOK FOR ME	Gardens, yards, parks, farm fields, and forests
WHAT I EAT	Worms, insects, and berries
PLANT THIS FOR ME	Berry-making trees and bushes, such as chokeberry, hawthorn, dogwood, sumac, and juniper
SOME THINGS TO KNOW ABOUT ME	American Robins are a kind of thrush. These birds are famous for their musical songs.
	When looking for food on the ground, a robin runs then stops suddenly, then runs and stops again.
	Robins nest in the lower half of small trees or bushes. They will use nest boxes and sometimes even build nests in hanging plants or over doorways. Their eggs are sky blue to blue-green.

BALTIMORE ORIOLE

....

Icterus galbula

....

Orange-and-black, medium-sized songbird with long legs and a pointy bill

WHERE DO I LIVE?

- Breeding
- Winter
- Migration

ID ME	**SIZE** 6½ to 7½ inches (17 to 19 cm) long **COLOR** Male has black head and back, bright orange chest, white bars on black wings; female colors slightly duller **VOICE** Lovely, flutelike string of notes
LOOK FOR ME	High up in the leafy trees of parks, backyards, edges of forests, and along rivers
WHAT I EAT	Insects, fruit, and nectar
PLANT THIS FOR ME	Fruit trees like crab apple and mulberry; berries like raspberry, blackberry, blueberry, and elderberry; flowers with nectar, such as trumpet creeper vines
SOME THINGS TO KNOW ABOUT ME	Baltimore Orioles are crazy about fruit and come to feeders filled with nectar, jelly, and split oranges. These acrobats hop between high branches and hang upside down from thin twigs to snag berries or chomp bugs. Baltimore Orioles weave soft, sacklike nests of grass and strips of bark that hang from branches like socks on a clothesline.

BIRDS

BLACK-BILLED MAGPIE

Pica hudsonia

Big, noisy, black-and-white bird with a long tail that travels in groups and perches on fence posts

WHERE DO I LIVE?

■ Year-round
■ Winter (scarce)

ID ME	SIZE 17½ to 23½ inches (45 to 60 cm) long
	COLOR Black head and bill, black-and-white body, iridescent black-blue wings and tail
	VOICE Loud raspy *jeeep-jeeep, wock-a-wock-a-wock-a, jeeep-jeeep*
LOOK FOR ME	Grasslands, parks, yards, farm fields, and pastures
WHAT I EAT	Grasshoppers, beetles, fruit, grain, small prey, and carrion
SOME THINGS TO KNOW ABOUT ME	Magpies come to platform and suet feeders. They're noisy, social, and fun to watch.
	Black-billed Magpies eat ticks off the backs of cattle and flip over dried cow dung looking for beetles.
	Nesting pairs build a domed nest of sticks around a mud cup lined with grass.

BLACK-CAPPED CHICKADEE

Poecile atricapillus

Fluffy and friendly, round little bird with a black cap and eyes, and a tiny black bill

WHERE DO I LIVE?

Year-round

ID ME	**SIZE** 4½ to 6 inches (12 to 15 cm) long **COLOR** Light gray body, black chin and top of head to eyes, white cheeks, light belly **VOICE** Two whistled notes *fee-bee*, as well as *chickadee-dee-dee*
LOOK FOR ME	Forests, parks, yards, and gardens
WHAT I EAT	Insects, seeds, and some berries
PLANT THIS FOR ME	Willow, alder, and birch trees grow up to become great nesting sites.
SOME THINGS TO KNOW ABOUT ME	Carolina Chickadees (*Poecile carolinensis*) are nearly identical to their black-capped cousins. If you live south of the middle of Missouri, Indiana, and Ohio, you're probably seeing a Carolina Chickadee. It sings a longer four-note call: *fee-bee-fee-bay*. Chickadees are cavity nesters and happily use nest boxes, especially with some wood shavings in them. And they love feeders! Black-capped Chickadees are social, curious, busy, "talkative" birds. Flocks have many different calls with specific meanings. For example, the alarm call to warn of a prowling cat is different than one for a hunting hawk.

BIRDS

BLACK-HEADED GROSBEAK

Pheucticus melanocephalus

Beefy songbird with thick, cone-shaped bill on a big head; flashes of yellow under the wings in flight

WHERE DO I LIVE?

Breeding
Year-round
Winter
Migration

ID ME	**SIZE** 7 to 7½ inches (18 to 19 cm) long **COLOR** Orange-brown body, black head, black-and-white wings. Females are duller in color all over and sometimes have darkish streaks on their sides. **VOICE** Rising and falling whistled song
LOOK FOR ME	Mountain forests; desert thickets by streams; yards and gardens with trees
WHAT I EAT	Beetles, spiders, snails, seeds, and fruit
PLANT THIS FOR ME	Fruit trees like crab apple and mulberry, and berry bushes like elderberry and serviceberry
SOME THINGS TO KNOW ABOUT ME	Grosbeaks eat sunflower seeds from feeders, as well as nectar set out for orioles. Unlike most birds, Black-headed Grosbeaks can eat wintering monarch butterflies even though the insects are full of toxins. Male grosbeaks court females with loud songs and fancy flying. To attract a mate, a singing male flaps his wings and flies up off his perch, spreading his wings and showing off his tail for 8 to 10 seconds before landing back where he started.

BLUE JAY

Cyanocitta cristata

Large, noisy songbird with black-and-blue wings, a white belly, and a crest of feathers on its head

WHERE DO I LIVE?

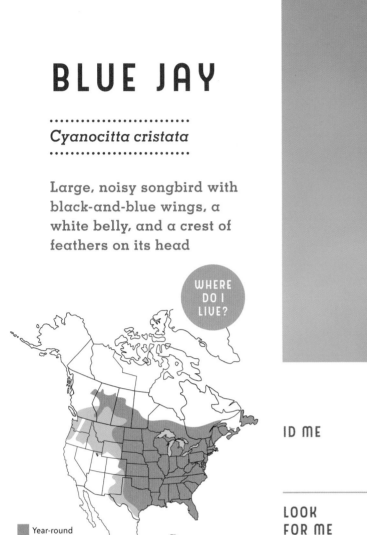

■ Year-round
■ Winter (scarce)

ID ME	SIZE 10 to 12 inches (25 to 30 cm) long
	COLOR Blue back with black, white, and blue markings, gray-white chest and belly
	VOICE Loud *jeer-jeer-jeer* as well as whistles; will mimic hawk calls
LOOK FOR ME	Forest edges, yards, and wherever oak trees grow
WHAT I EAT	Acorns, insects, nuts, seeds, fruits, small prey, and carrion
PLANT THIS FOR ME	Oak trees
SOME THINGS TO KNOW ABOUT ME	Blue Jays hold their crests high when being aggressive or feeling threatened. Crests are lowered during mellow family time.
	Jays like platform feeders with peanuts, suet, and sunflower seeds. They visit and drink from birdbaths, too.
	Blue Jays can learn to mimic cats, hawks, and even human speech sounds.

BIRDS

CAROLINA WREN

Thryothorus ludovicianus

Small, hopping, brown bird with upright tail

WHERE DO I LIVE?

■ Year-round
■ Year-round (scarce)

ID ME	**SIZE** 4½ to 5½ inches (12 to 14 cm) long **COLOR** Reddish brown, tan chest, long white eyebrow stripe, down-curving beak **VOICE** Very loud *teakettle-teakettle-teakettle* call
LOOK FOR ME	Woods and areas of thick overgrown thickets, vines, and bushes
WHAT I EAT	Spiders; insects including caterpillars, crickets, beetles; suet from feeders
PLANT THIS FOR ME	Thick bushes and vines, dense trees
SOME THINGS TO KNOW ABOUT ME	These birds are fun to watch. They hop and flit from log to leaf pile and climb tree trunks looking for bugs. Carolina Wrens are cavity nesters, making nests 3 to 6 feet (1 to 1.8 m) off the ground in tree holes or nest boxes. Nesting wrens like to get into sheds and other safe spaces. You might find a flower pot or work boot stuffed with leaves and grasses!

COMMON GRACKLE

Quiscalus quiscula

Lean, long-tailed blackbird with a shiny body and gold-rimmed eyes, often seen pecking at lawns

WHERE DO I LIVE?

Breeding
Year-round
Winter

ID ME	**SIZE** 11 to 13½ inches (28 to 34 cm) long
	COLOR Black body; head and neck shiny dark blue. Females are a bit smaller and not as glossy.
	VOICE Loud *readle-eak, readle-eak* along with squeaks and whistles
LOOK FOR ME	Farm fields, pastures, parks, backyards, marshes, forest edges, and meadows
WHAT I EAT	Seeds from trees, flowers, and grains; some insects; other small prey
SOME THINGS TO KNOW ABOUT ME	Grackles come to feeders with mixed seed and are happy to eat what's spilled on the ground below.
	Noisy and social, these birds like hanging out in big chatty flocks with other birds.
	When grackles are looking for mates in spring, they playfully chase each other in flight.

DARK-EYED JUNCO

Junco hyemalis

Little, round bird with dark hood snacking on seed in flocks beneath feeders, especially in winter

WHERE DO I LIVE?

- Breeding
- Year-round
- Winter

ID ME	SIZE 5½ to 6 inches (14 to 16 cm) long
	COLOR Gray with a dark hood, pale belly, pink bill
	VOICE Long, even musical trill
LOOK FOR ME	Fields, parks, gardens, and woodlands during winter and while migrating; forests during summer breeding season
WHAT I EAT	Seeds in winter; some insects in breeding season
PLANT THIS FOR ME	Seed-making native sedges and grasses, native sunflowers
SOME THINGS TO KNOW ABOUT ME	Juncos live all across North America, from Mexico to Canada and Massachusetts to California.
	In winter climates, juncos often show up beneath feeders as the cold sets in.
	Juncos in western North America are brown with a dark hood and white belly.

EASTERN BLUEBIRD

Sialia sialis

Small, plump bird with blue back and large eyes hanging out on posts and low branches

WHERE DO I LIVE?

Breeding

Year-round

Winter

ID ME	SIZE 6 to 8 inches (16 to 21 cm) long
	COLOR Bright blue on top, rusty red chest, buff belly; female coloring is duller
	VOICE Soft, low-pitched warble: *tu-a-wee*
LOOK FOR ME	Meadows, grasslands, and other open areas surrounded by trees
WHAT I EAT	Insects such as caterpillars, crickets, and beetles caught on the ground in summer; fruits and berries in winter
PLANT THIS FOR ME	Fruiting plants and trees: mistletoe, sumac, blueberry, black cherry, tupelo, wild holly, dogwood, hackberry, pokeweed, and juniper
SOME THINGS TO KNOW ABOUT ME	Eastern Bluebirds nest in tree cavities and old woodpecker holes.
	Bluebirds like smaller nest boxes no bigger than 4 inches (10 cm) wide with holes around 1.8 inches (4.5 cm) across.
	Western Bluebirds (*Sialia mexicana*) have a blue throat. Eastern Bluebirds have a rusty throat. And Mountain Bluebirds (*Sialia currucoides*) are blue all over.

GREAT HORNED OWL

Bubo virginianus

Giant, husky owl with earlike tufts and big yellow eyes

WHERE DO I LIVE?

Year-round

Year-round (scarce)

ID ME	SIZE 18 to 25 inches (46 to 63 cm) long
	COLOR Splotchy gray to red-brown with bars and patterns
	VOICE Deep, soft, stuttering: *hoo-h'HOO-hoo-hoo*
LOOK FOR ME	Woods, swamps, farmland, deserts, and suburbs
WHAT I EAT	Rabbits, woodchucks, bats, skunks, ducks, crows, and prey of all kinds
SOME THINGS TO KNOW ABOUT ME	Great Horned Owls sometimes nest in cavities. More often they take over abandoned stick nests built by other large birds or nest in forked branches. These owls aren't fussy eaters, hunting anything from scorpions to skunks. They are fierce hunters who will tackle prey larger than themselves. This owl's short, wide wings are perfect for flying through thick forests. Its soft feathers make no noise in flight. Like all owls, it has excellent night vision and keen hearing.

HOUSE FINCH

Haemorhous mexicanus

Small, streaked bird with red coloring around face and upper chest and a thick bill

WHERE DO I LIVE?

■ Year-round

ID ME	**SIZE** 5 to 5½ inches (13 to 14 cm) long
	COLOR Gray-brown with streaks, red on face, head, and chest. Females have the same streaky patterns without the red.
	VOICE Long, warbling song of short, twittering notes
LOOK FOR ME	Lawns, parks, farmland, grasslands, deserts, and forests
WHAT I EAT	Seeds, buds, and fruit
PLANT THIS FOR ME	Knotweed; native thistle, sedges, and sunflowers; mulberry
SOME THINGS TO KNOW ABOUT ME	House Finches are native to western North America. But in 1940 a number of caged House Finches were set loose in New York. The birds reproduced and spread across the East.
	House Finches travel in little flocks visiting feeders, especially ones with black oil sunflower seeds.
	Males will feed females during courtship to show their parenting skills. Like a hungry chick, a female pecks on the male's bill and flaps her wings to ask for food.

BIRDS

NORTHERN CARDINAL

Cardinalis cardinalis

Large, red songbird with thick bill and tall crest of feathers on head

WHERE DO I LIVE?

■ Year-round

ID ME	**SIZE** 8 to 9 inches (21 to 23 cm) long **COLOR** Males: bright red body, black face, red bill. Females are tawny light brown with reddish tinge on wings, tail, and head; red bill **VOICE** Loud string of whistles that end in a trill: *bird-EE, bird-EE, bird-EE*
LOOK FOR ME	Backyards, overgrown fields, and forest edges with high perches
WHAT I EAT	Seeds, fruits and berries, and some insects
PLANT THIS FOR ME	Fruiting plants such as dogwood, mulberry, hackberry, blackberry, sumac, tulip tree, and wild grape; grasses and sedges
SOME THINGS TO KNOW ABOUT ME	Cardinals feed in low branches or on the ground but can be seen and heard singing from the tops of trees or on telephone poles. Courting pairs build hidden nests in tangles of vines or where two small branches fork. Cardinals come to feeders and especially like sunflower seeds.

NORTHERN FLICKER

....

Colaptes auratus

....

Big, grayish brown woodpecker with an easy-to-see white rump patch when flying

WHERE DO I LIVE?

Breeding

Year-round

Winter

ID ME	**SIZE** 11 to 12 inches (28 to 31 cm) long **COLOR** Grayish brown, dark spots on front, dark bars on wings. Feathers on the undersides of the wings and tail are bright yellow in eastern flickers and red in western flickers. **VOICE** Loud rolling rattle: *ka-ka-ka-ka-ka* and rapid drumming on trees
LOOK FOR ME	Parks, yards, woodlands, mountain forests, and trees along open country
WHAT I EAT	Insects like ants and beetles in summer; berries and seeds in winter
PLANT THIS FOR ME	Berry and seed plants like dogwood, sumac, wild cherry, grape, bayberry, hackberry, and elderberry; sunflowers; and native thistle
SOME THINGS TO KNOW ABOUT ME	Like most woodpeckers, flickers fly in an up-and-down pattern of flapping then gliding, flapping then gliding. Unlike most woodpeckers, Northern Flickers look for food on the ground and will perch on branches instead of clinging to tree trunks. Flickers competing over territory or mates will face off on a branch and use their bills in fencinglike duels. *En garde!*

BIRDS

RED-TAILED HAWK

Buteo jamaicensis

Big hawk sitting on utility poles along roads and highways or circling over fields and open areas

WHERE DO I LIVE?

- Breeding
- Year-round

ID ME	SIZE 17½ to 22 inches (45 to 56 cm) long
	COLOR Brown on top, cream below, rusty tail
	VOICE Loud harsh scream: *kee-eeeee-arr*
LOOK FOR ME	Deserts, grasslands, fields and pastures, roadsides, and open edges of woodlands
WHAT I EAT	Rabbits, voles, mice, squirrels, and other small mammals; snakes, carrion, and sometimes birds
SOME THINGS TO KNOW ABOUT ME	Red-tailed Hawks are armed with a flesh-tearing, hooked beak and sharp, prey-grabbing claws called talons.
	If you see a hawk going after birds at a feeder, it's probably a smaller Cooper's or Sharp-shinned Hawk, not a big Red-tailed.
	Red-tailed Hawk pairs build nests of sticks in the tops of tall trees, or on cliff ledges. A nest that is added to over many seasons can grow as tall as a person! They line the inside with strips of bark, leaves, and grasses.

RUBY-THROATED HUMMINGBIRD

Archilochus colubris

Humming sound and flash of bright green

WHERE DO I LIVE?

- Breeding
- Winter
- Migration

ID ME	**SIZE** 3 to 3½ inches (7 to 9 cm) long **COLOR** Shiny green with white chest; males have red throats, females have white throats **VOICE** Quick, squeaky *chip* (plus the hum of wings)
LOOK FOR ME	Flower-filled gardens, parks, meadows, edges of woods, and nectar feeders
WHAT I EAT	Flower nectar and small insects like gnats and aphids
PLANT THIS FOR ME	Plants and bushes with red, orange, and tube-shaped flowers: trumpet creeper, bee balm, jewelweed, cardinal flower, bergamot, columbine, fire pink, red buckeye, and phlox
SOME THINGS TO KNOW ABOUT ME	These little birds lay jelly-bean-sized eggs in tiny nests made of lichen, dandelion fuzz, and sticky spider webs. They migrate hundreds of miles to spend winters in Central America, so they need lots of food for the long journey. High-energy hummingbirds beat their wings 53 times a second and can hover and fly backward like a helicopter.

RUFOUS HUMMINGBIRD

....

Selasphorus rufus

....

Humming sound and flash of shiny orange

WHERE DO I LIVE?

Breeding

Winter

Migration

ID ME	**SIZE** 3 to 3½ inches (7 to 9 cm) long **COLOR** Males are bright orange with white chest and shiny red throat. Females have greenish backs, cream-colored breasts, and rusty patches on sides, tails, and throat. **VOICE** Short *chip* (and hum of wings)
LOOK FOR ME	Flower-filled parks and backyards, as well as nectar feeders
WHAT I EAT	Sweet flower nectar and small insects like gnats and aphids
PLANT THIS FOR ME	Colorful, tube-shaped flowers like columbine, Indian paintbrush, mint, fireweed, and larkspur
SOME THINGS TO KNOW ABOUT ME	Rufous Hummingbirds are feisty and aggressive. They'll chase away other hummingbirds from feeders. They are tiny travelers of long distances, some flying from Alaska to Mexico (and back) each year. Hummingbirds catch small insects out of the air or pluck them off spiderwebs, leaves, or tree bark. Bugs are an important source of protein.

SONG SPARROW

·······················

Melospiza melodia

·······················

Plump sparrow covered in brown streaks proudly singing out in the open

WHERE DO I LIVE?

- Breeding
- Year-round
- Winter

ID ME	**SIZE** 4½ to 6½ inches (12 to 17 cm) long
	COLOR Streaks of brown to red-brown on white to gray body
	VOICE Long, loud, stuttering song that ends with a buzz or trill
LOOK FOR ME	Overgrown fields, yards, and forest and marsh edges
WHAT I EAT	Insects, spiders, snails, and worms, as well as seeds and fruit
PLANT THIS FOR ME	Native clovers, ragweed, sunflowers, and berries
SOME THINGS TO KNOW ABOUT ME	Song Sparrows are common across most of North America but look different depending on where they live. Some are bigger and darker, others are paler and smaller.
	Like many songbirds, males sing both to attract a mate and to defend territory. That pretty song is saying, "Get out of here! NOW!"
	Courting Song Sparrows fly on fluttery wings as a pair with tails up and legs dangling down.

BIRDS

TUFTED TITMOUSE

Baeolophus bicolor

Small, gray bird with pointy crest and big black eyes cracking open seeds on a twiggy perch

WHERE DO I LIVE?

Year-round

ID ME	SIZE 5½ to 6 inches (14 to 16 cm) long
	COLOR Shiny gray on top, white below, some rust along the sides, black patch between black eyes, black bill
	VOICE Whistled *peter-peter-peter*
LOOK FOR ME	Woods with lots of tall trees, orchards, parks, and yards
WHAT I EAT	Spiders; insects like caterpillars, beetles, and ants; seeds, nuts, and berries
PLANT THIS FOR ME	Nut trees like beech and oak trees for acorns; native sunflowers
SOME THINGS TO KNOW ABOUT ME	Tufted Titmice love to eat sunflower seeds, suet, and peanut butter from feeders.

Titmice nest in holes found in dead trees, old woodpecker nest holes, as well as nest boxes. They can't dig out their own nest holes.

The Black-crested Titmouse (*Baeolophus atricristatus*) has a white forehead and jet black crest. It lives in Texas and northeastern Mexico. |

TURKEY VULTURE

Cathartes aura

Huge, dark bird soaring high in the sky with V-shaped wings that end in fingerlike feathers

WHERE DO I LIVE?

Breeding

Year-round

ID ME	SIZE 25 to 32 inches (64 to 81 cm) long
	COLOR Dark brown to black, red head, whitish beak
	VOICE Hisses
LOOK FOR ME	Farm fields, roadsides, and open country
WHAT I EAT	Carrion (dead animals)
SOME THINGS TO KNOW ABOUT ME	Unlike most birds, vultures have an excellent sense of smell. They use it to track down their dead dinner.
	Besides high up in the sky, a good place to spot vultures is along roadsides and highways, especially near roadkill.
	Turkey vultures can't sing, but they do hiss to scare away other scavengers.

BIRDS

ANTS

3

ATTRACT
BUTTERFLIES, BEES
& OTHER
BUGGY BUDDIES

MOTH

A SPIDER WEAVING A WEB or a cicada leaving its shell are awesome sights. Bugs provide some of the best backyard wildlife watching around. Why? They're easy to find and live almost everywhere. Even a grassy strip between sidewalk and street is home to ants and worms, pill bugs, and beetles.

INSECTS, SPIDERS, SLUGS, WORMS, and other small creatures are what backyard habitats are built on. It's true! Insects are important pollinators. Bees and butterflies fertilize plants by transferring grains of pollen between flowers. Without pollination, flowers can't make seeds, nuts, or fruits. The food that wildlife (and people, too!) needs to survive depends on pollen-carrying bugs.

INSECTS ARE INVERTEBRATES, MEANING THEY DON'T HAVE BONES.

HEALTHY SOIL ALSO COMES from the work of worms, slugs, and insects. They break down dead leaves and transform animal remains into nutrients. And think of all the frogs, birds, shrews, snakes, and other backyard critters that eat bugs and slugs, spiders and insects.

DRAGONFLY

69

CRICKET

WHAT IS AN INSECT?

AN INSECT IS A SIX-LEGGED ANIMAL that lives as both a larva and an adult — and sometimes another step in between. A good insect habitat has food for each stage of an insect's life. While bumblebee larvae eat pollen, the adults need nectar to sip. Butterfly eggs hatch into larvae called caterpillars. Many caterpillars only eat the leaves of certain plants, called host plants. The host plant for clouded sulphur butterflies is clover, for example.

INSECTS LAY EGGS in all sorts of places. Dragonflies lay eggs in water. Many bugs stash eggs in soil. Other insects need good shelter and nest sites to build their hives or colonies. Ants, wasps, and bees are famous makers of nests where larvae are fed and raised.

MANY NATIVE BEES ARE IN TROUBLE because of pesticides and loss of habitat. Piles of brush, dead trees, chemical-free lawns, and native plants can make a big difference to their survival.

IT'S NOT HARD TO HELP OUT INSECTS AND BUGS. THEY'RE ALL AROUND YOU!

CATERPILLAR

BIO BASICS

Insects

beetles, flies, butterflies, moths, dragonflies, crickets, grasshoppers, bees, wasps, ants, and more

Here are some of the invertebrates (creatures without a skeleton) that might live near you.

Myriapods

centipedes and millipedes

Arachnids

spiders, harvestmen, ticks, mites

Mollusks

snails and slugs

Crustaceans

pillbugs and woodlice

Worms

earthworms and leeches

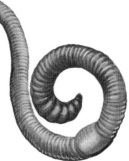

MATERIALS

30–40 paper drinking straws or paper, pencil, and tape to make your own

Scissors

Metal soup or bean can, washed and dried

Wood glue

2 or 3 cardboard toilet tissue tubes

Twine or string

BE HOME, BEES

Honeybees are not native to North America. They were one of the many farmed animals brought by European settlers. Most native North American bees are smaller than honeybees and nest in holes and crevices, not hives. This whole-lot-of-holes native bee home is a bee-utiful part of any backyard habitat.

HOW TO

1 If you don't have paper straws, you can make your own. Roll a small sheet of paper around a pencil to make a tube. Tape the edge of the paper and pull out the pencil. Repeat 30 to 40 times!

2 Cut the paper straws or tubes to be about ½ inch (1.3 cm) shorter than the can.

3 Squeeze a thick coating of glue into the bottom of the empty can and spread it around. Push two or three cardboard tubes into the can.

4 Fill the cardboard tubes and the area around them with the paper straws. Push all the pieces down to be sure they are held in place by the glue. Let dry.

5 Tie twine or string about one-third of the way from the bottom of the can, so that the open end will tilt downward to keep out rain. Hang your bee house from a tree branch, fence, or porch railing.

BEE BLOCKS AND BOXES

Know someone handy with a drill? A hole-filled wood block or log makes a great native bee nest.

Want to make a wooden bumblebee box? Look for the web address on page 175.

The holes can be redrilled several times to provide a home for generations of bees.

The holes should be ⅛ to ⅜ inch (0.3 to 0.9 cm) wide and at least 5 inches (13 cm) deep.

POLLINATOR GARDEN

Butterflies, native bees, and other pollinators are in trouble across North America. You can help by planting a pollinator garden. Native bees need nectar-making flowers and places to nest. Adult butterflies need nectar, too, and their caterpillars eat particular host plants.

Planting different native flowers that bloom in sunny spots throughout the growing season is a big help. So are host plants like milkweeds and clovers. And skip the pesticides!

THISTLE

SUNFLOWER

WILD BERGAMOT

LEAVE IT BEE

When beetles burrow in dead trees and branches they leave behind tunnels. Wood-nesting bees lay eggs in these tunnels. Leaving dead trees and piles of branches be is good for native bees.

To learn which pollinator plants and butterfly host plants are best for your area, look for the web addresses on pages 174–75.

75

MATERIALS

Scissors

½-inch (1.3 cm) thick sponge

Opened paper clip

Empty ketchup, mustard, or other squeeze bottle with lid

String or twine

Stickers, indelible markers, or fake flowers and glue for decoration

Small bowl and spoon

Extrafine sugar

Water

Soy sauce

Sand and rocks

Shallow dish, container, tray, or saucer

BUTTERFLY CAFÉ

Nothing is nicer than spotting a butterfly sipping nectar or drinking from a puddle. These beautiful insects are important pollinators. So serve them up a snack to help them do their job!

HOW TO

1 Cut a piece of sponge about ½ x 3 inches (2 x 7 cm). Use the paper clip to poke the sponge through the hole of the squeeze bottle lid from the bottom. Pull about an inch of sponge through to stick out the top.

2 Tie string around the bottle to make a hanger. Decorate the bottle with bright stickers or colorfast fake flowers. The colors red, yellow, orange, pink, and purple bring in butterflies.

4 Hang the feeder with the sponge tip facing down. Make a drinking puddle by adding a handful of sand and a few rocks to a shallow dish. Set the dish above the ground near the feeder. Note how many different types of butterflies you attract!

3 Mix 4 teaspoons (20 mL) of sugar into 1 cup (240 mL) of hot tap water until dissolved. Add 2 or 3 drops of soy sauce. Pour the nectar into the squeeze bottle and screw the lid on tightly.

BEE BALM

······ HELPFUL PREDATOR PLANTS

Pest-eating insects like mantises, ladybugs, robber flies, and predatory wasps are meat eaters. But many also eat nectar or pollen and all need plants for cover and nesting spaces. Plant these to attract helper bugs: asters, goldenrod, native sunflowers, cup plant, ironweed, bee balm, coneflowers, golden alexanders, meadowsweet. Use the native plant websites on pages 174–75 to help you choose what grows best where you live.

GOLDENROD

ASTER

WILD NOTES
BUGS & INSECTS

Wondering what else to take note of?

- Keep a butterfly calendar, writing down the dates when different migrators show up, especially monarchs.

- Use the Field Guide pages to figure out which insects and other bugs you're seeing. Keep a list!

- Investigate insect singers. Keep a list of the insect sounds you hear and who might be making them. Use the bug websites on page 174 for help.

- Project updates are important! What's working, what's not? Any ideas for improvements?

····· SHARE THE LOVE

Many butterfly-attracting plants bring in hummingbirds, too. Both are nectar lovers!

BANANA SLUG

Ariolimax columbianus

Huge slug that looks like an overripe banana

WHERE DO I LIVE?

ID ME	SIZE up to 10 inches (25 cm) long
	COLOR Yellow with brown spots
LOOK FOR ME	Moist forest floors of the Pacific Coast
WHAT I EAT	Leaves, dead plants, fungi, animal droppings
SOME THINGS TO KNOW ABOUT ME	Banana slugs are native to North America. Many common slugs — especially the garden pest kinds — were introduced from Europe.
	All slugs have two pairs of tentacles on their heads. One is for sensing light, the other for smell.
	Slugs and land snails are decomposers, breaking down dead plants and animals into soil.

CECROPIA MOTH

··························
Hyalophora cecropia
··························

Large, colorful moth with striped body, patterned wings, and feathery antennae

WHERE DO I LIVE?

ID ME	**WINGSPAN** 4½ to 6 inches (11 to 15 cm) wide **COLOR** Gray to black wings with red-and-white bands, crescents, and eyespots
LOOK FOR ME	Neighborhoods, forests, parks, and woodlands
WHAT I EAT	Caterpillars eat tree leaves; adults don't eat at all.
PLANT THIS FOR ME	Trees like ash, birch, box elder, elm, maple, poplar, dogwood, and willow
SOME THINGS TO KNOW ABOUT ME	The caterpillars are green with spiky red, yellow, or blue knobs all over their bodies. Moths are nocturnal, so look for them at night. This is the largest native moth species in North America.

BUGS & INSECTS

COMMON WALKING STICK

Diapheromera femorata

Slow-moving, long, thin insect that looks exactly like a twig

WHERE DO I LIVE?

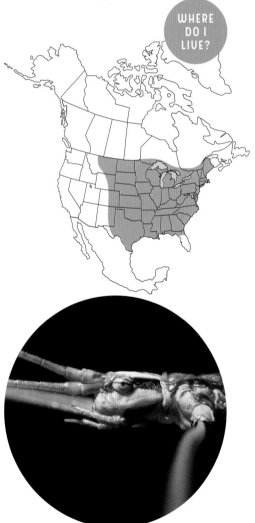

ID ME	SIZE 2½ to 4 inches (6 to 10 cm) long
	COLOR Green to brown
LOOK FOR ME	Leafy forests
WHAT I EAT	Tree and bush leaves
PLANT THIS FOR ME	Hardwood trees like oak, hickory, wild cherry, and black locust
SOME THINGS TO KNOW ABOUT ME	There are some half-dozen kinds of walking sticks, or stick insects, in North America. Each is perfectly camouflaged in its prairie, forest, or scrubland home.
	Common walking sticks change color from green to brown to green.
	Walking sticks are wingless, but if they lose a leg, they can grow a new one.

DIFFERENTIAL GRASSHOPPER

Melanoplus differentialis

Big, shiny grasshopper with huge eyes and V-shaped markings on back legs

WHERE DO I LIVE?

ID ME	SIZE 1 to 2 inches (3 to 5 cm) long
	COLOR Light brown with black markings
	VOICE *Buzz-buzz*
LOOK FOR ME	Open woods, fields, and meadows
WHAT I EAT	Plant and grass leaves
SOME THINGS TO KNOW ABOUT ME	Grasshoppers and crickets make noises by rubbing wings together or rubbing a hind leg against a wing.
	Differential grasshoppers are great jumpers thanks to powerful hind legs.
	The females lay egg pods in the soil. Each pod has up to eight eggs inside it.

GARDEN SPIDER

Argiope aurantia

Large yellow-and-black spider with long legs on a distinctive web

WHERE DO I LIVE?

EGG SACK

ID ME	**BODY** Females ¾ to 1 inch (20 to 28 mm) long; males one-third the size of females.
	COLOR Yellow-and-black markings on egg-shaped lower body; black legs with yellow to orange sections. Males are paler.
LOOK FOR ME	Gardens, meadows, parks, and riverside areas
WHAT I EAT	Flying insects
SOME THINGS TO KNOW ABOUT ME	In fall, females weave a silky sheet, lay hundreds of eggs on it, cover the eggs in more silk, and then wrap it up into a ball-shaped egg sack. The spiderlings hatch out and leave the sack in spring.
	The garden spider weaves a thick, zigzag ribbon into the web's center.
	The male hangs out on the edge of the web. The female waits for snagged prey in the web's center.

GREEN DARNER DRAGONFLY

....
Anax junius
....

Large, fast-moving dragonfly
with clear wings

WHERE DO I LIVE?

ID ME	BODY 2½ to 3 inches (7 to 8 cm) long
	COLOR Bluish head, green chest with clear wings, and a purplish or blue lower body
LOOK FOR ME	Ponds, streams, and fields
WHAT I EAT	LARVAE water insects, small fish, tadpoles
	ADULTS flying insects
SOME THINGS TO KNOW ABOUT ME	This is one of the biggest dragonflies in North America.
	Females lay eggs in water, which hatch into larvae that are called nymphs. These young dragonflies are water-living hunters, eating insects, little fish, and even tadpoles. Green darners live as nymphs for a year before changing into adult dragonflies.
	The green darner is one of the few kinds of dragonflies that migrate south for the winter.

BUGS & INSECTS

GROVE SNAIL

....

Cepaea nemoralis

....

A small land snail with a shiny shell

WHERE DO I LIVE?

Map based on reported sightings; actual range likely greater.

ID ME	SHELL about 1 inch (25 mm) wide
	COLOR Yellow to brown shell with dark bands
LOOK FOR ME	Under logs and rocks or in hollow trees in parks or neighborhoods
WHAT I EAT	Dead plants
SOME THINGS TO KNOW ABOUT ME	Female land snails lay round, rubbery eggs in damp places.
	Land snails and slugs are mollusks. That makes them related to octopuses, squid, and clams.
	Many common land snails (including this one) aren't native to North America. They were introduced from Europe and are garden pests.

HARVESTMAN

Phalangium opilio

Small, oval body with eight extralong, threadlike legs

WHERE DO I LIVE?

ID ME	BODY $\frac{1}{10}$ to $\frac{1}{5}$ inch (4 to 6 mm) long
	COLOR Red-brown, dark brown legs
LOOK FOR ME	Neighborhoods, woods, forests, mountains, farm fields, meadows, and parks
WHAT I EAT	Rotting leaves and plants; aphids, mites, and other small bugs
SOME THINGS TO KNOW ABOUT ME	Harvestmen like shady places such as leaf piles and wood stacks.
	They might have eight legs, but harvestmen are not spiders. They're a different kind of arachnid, the group of eight-legged invertebrates that includes spiders, scorpions, mites, and ticks.
	Many people call these common creatures *daddy longlegs*. Do you?

BUGS & INSECTS

KATYDID

....

Pterophylla camellifolia

....

Green, round-bodied insect with wings that look like leaves

WHERE DO I LIVE?

ID ME	SIZE 2 inches (5 cm) long
	COLOR Bright green
	VOICE *katy-DID, katy-DID*
LOOK FOR ME	Forests, woods, and neighborhoods
WHAT I EAT	Leaves of trees
PLANT THIS FOR ME	Leafy trees
SOME THINGS TO KNOW ABOUT ME	Katydids are much easier to hear than see. They live in the crown, or top, of trees. Plus, katydids are perfectly camouflaged and blend in with the leaves. Fork-tailed bush katydids (*Scudderia furcata*) live across the United States. They have longer, narrower bodies and bigger legs. Katydids are relatives of grasshoppers and crickets.

LADY BEETLE

....
Coccinellidae
....

Small, oval beetle with black dots on red wing covers

WHERE DO I LIVE?

ASIAN LADY BEETLE

ID ME	BODY about ⅕ inch (5 mm) long
	COLOR Black spots on red wing covers
LOOK FOR ME	Woods, meadows, gardens, and parks
WHAT I EAT	Aphids, mites, and other small bugs

SOME THINGS TO KNOW ABOUT ME

These small beetles come in many colors and patterns. Some people call them ladybugs or ladybird beetles.

There are more than one hundred kinds of lady beetles in the United States. The similar-looking Asian lady beetle is a different species and is considered invasive.

Lady beetles are a gardener's friend. They eat plant pests, like aphids.

Females lay clusters of 15 to 30 bright yellow eggs on plants during spring and summer.

BUGS & INSECTS

MASON BEE

Osmia spp.

Small, hairy, flylike black bee on flowers

WHERE DO I LIVE?

ID ME	SIZE about ½ inch (1 cm) long COLOR Black
LOOK FOR ME	Meadows and open woods, gardens, and parks
WHAT I EAT	LARVAE nectar and pollen ADULTS pollen
PLANT THIS FOR ME	Nectar-filled flowers like native sunflowers and asters
SOME THINGS TO KNOW ABOUT ME	Mason bees, like many native North American bees, are solitary. They live alone instead of in colonies with workers and a queen. These native bees are important pollinators. Their hairy fuzz is perfect for trapping and transporting pollen between flowers. Mason bees use mud to build nests in small crevices and corners. You can help by giving them places to nest (see pages 72–73).

MONARCH BUTTERFLY

....
Danaus plexippus
....

Large butterfly with deep-orange wings that have black veins and black borders with white dots

WHERE DO I LIVE?

ID ME	WINGSPAN 3 to 4 inches (8 to 10 cm) wide
	COLOR Orange and black
LOOK FOR ME	Fields, gardens, and roadsides
WHAT I EAT	CATERPILLARS milkweed leaves
	ADULTS nectar from flowers
PLANT THIS FOR ME	Milkweed, nectar-filled flowers like native sunflowers and asters

SOME THINGS TO KNOW ABOUT ME

Like all butterflies, monarchs hatch out as caterpillar larvae. They eat and grow, eventually attaching themselves to a stem and transforming into a chrysalis. Weeks later, an adult butterfly breaks out of the cocoon.

Monarch caterpillars feed on the gluey, sap-filled leaves of milkweed, soaking up its toxic chemicals. Most birds won't eat bad-tasting monarchs.

Every fall, monarchs living east of the Rocky Mountains migrate to Mexico for the winter. In spring, they head back north, laying eggs along the way then dying at the end of the season. Somehow their offspring know how to find their way to Mexico to repeat the cycle.

BUGS & INSECTS

PAPER WASP

·················
Polistes spp.
·················

Slender wasp with long wings on a paper nest

WHERE DO I LIVE?

ID ME	BODY ½ to 1 inch (10 to 25 mm) long
	COLOR Red-brown body with yellow and black markings
	SOUND *Buzzzzzz* from wings
LOOK FOR ME	Buildings, gardens, parks, and open woods
WHAT I EAT	Nectar, sap, fruit juices; insects to feed larvae
SOME THINGS TO KNOW ABOUT ME	There are about two dozen kinds of paper wasps (*Polistes*) in the United States and Canada. Most can deliver a painful sting, so don't touch!
	Wasps make paper by chewing up plant stems and wood and mixing it with spit. They use the paper to build a nest of six-sided open cells.
	The queen lays eggs in the paper cells. Once the eggs hatch, workers feed the larvae chewed-up insects.

PERIODICAL CICADA

....

Magicicada spp.

....

Loud, large, black flying insect
with big, round, red eyes

WHERE DO I LIVE?

ID ME	
	SIZE 1 to 1½ inches (2.5 to 3.8 cm) long
	COLOR Black body, clear wings with orangish veins, red eyes
	VOICE Loud drone: *weeee-whoa, weeee-whoa*
LOOK FOR ME	Suburbs, parks, and forests with leafy trees
WHAT I EAT	Sap from underground roots, tree sap
PLANT THIS FOR ME	Leafy trees like oaks
SOME THINGS TO KNOW ABOUT ME	Cicada larvae hatch out of eggs laid in slits cut into tree stems by adult females. The nymphs climb trees when they're ready to become adults. The winged adults leave behind their shells on tree bark and fence posts.
	Periodical cicadas stay underground for 13 to 17 years (depending on the species). At the same time, they change into adults in huge broods. Then the thousands of males all sing together to attract mates. It makes for a noisy summer!
	There are also annual cicadas that hatch out in just 1 year. They don't have red eyes.

ANNUAL CICADA

BUGS & INSECTS

PILL BUG

Armadillidium vulgare

Tiny, oval bug with 14 legs and armorlike plates

WHERE DO I LIVE?

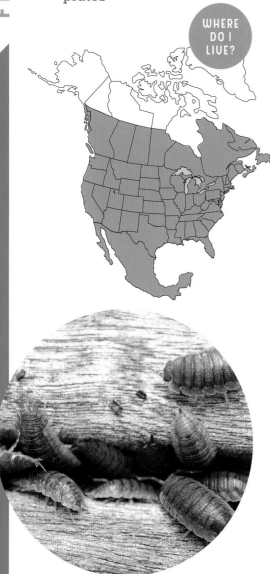

ID ME	SIZE ½ to 1 inch (1 to 2 cm) long
	COLOR Gray to black
LOOK FOR ME	Damp, dark places like underneath logs, rocks, or leaves
WHAT I EAT	Decaying plants and animals
SOME THINGS TO KNOW ABOUT ME	Whether you call them pill bugs, sow bugs, or roly-polies, these creatures aren't insects at all. They're crustaceans and related to lobsters, crayfish, and shrimp.
	Pill bugs roll themselves up into a ball to defend themselves.
	Pill bugs and their crustacean cousin the wood louse (*Porcellio scaber*) arrived with European settlers and their imported crops and livestock.

PRAYING MANTIS

Mantis religiosa

Large, slender insect with folded "praying" front legs and a triangle-shaped head

WHERE DO I LIVE?

ID ME	SIZE 2 to 2½ inches (5 to 6 cm) long
	COLOR Green to light brown
LOOK FOR ME	Fields, gardens, roadsides, and parks
WHAT I EAT	Insects and spiders

SOME THINGS TO KNOW ABOUT ME

Mantids are the only insects whose necks turn far enough to let them look over their own shoulders, which comes in handy for hunting.

All mantids are "preying" mantids! They are ambush hunters that wait for prey to get close enough to catch.

Gardeners use mantids to control plant pests. Mantids imported from Europe and Asia are now common all across North America. Carolina mantids (*Stagmomantis carolina*) are native to the southern half of the United States.

BUGS & INSECTS

TIGER SWALLOWTAIL

Papilio glaucus

Large, yellow-and-black butterfly with tail-like extensions on its back wings

WHERE DO I LIVE?

ID ME	**WINGSPAN** 3½ to 6 inches (9 to 16 cm) wide **COLOR** Yellow wings with black stripes and shiny blue hindwing splotches
LOOK FOR ME	Forests, meadows, parks, gardens, and roadsides
WHAT I EAT	**CATERPILLARS** tree leaves **ADULTS** nectar
PLANT THIS FOR ME	Poplar and black cherry trees for caterpillars, nectar-filled flowers like native sunflowers for adults
SOME THINGS TO KNOW ABOUT ME	A tiger swallowtail caterpillar is dark green with two large, eye-shaped spots on a swollen area of its back behind the head. Swallowtail butterflies have "tails" on their back wings that look a bit like the forked, pointy tails on birds called swallows. Tiger swallowtail butterflies are high fliers, cruising above treetops more than 160 feet (50 m) above the ground.

YELLOW BUMBLEBEE

Bombus fervidus

Large, fuzzy, black-and-yellow bee with dark wings, seen buzzing around flowers

WHERE DO I LIVE?

ID ME	BODY about ½ inch (13 mm) long
	COLOR Black head, yellow body with black bands, dark brown wings
	SOUND *Buzzzzz* from wings
LOOK FOR ME	Gardens, meadows, parks, roadsides, and open woods
WHAT I EAT	Nectar from flowers
PLANT THIS FOR ME	Flowers like black-eyed Susans, native sunflowers, milkweed, goldenrod, jewelweed, asters
SOME THINGS TO KNOW ABOUT ME	Yellow bumblebees live in colonies of 50 to 125 workers and one slightly larger queen. Watch with caution! Bumblebees sting.
	The similar-looking American bumblebee (*Bombus pensylvanicus*) lives farther south, but not as far west.
	There are more than 40 kinds of bumblebees in North America. Many of their populations are declining because of habitat loss and pesticides. They are important pollinators.

BUGS & INSECTS

PAINTED TURTLE

4

BRING IN
TURTLES, LIZARDS
& OTHER SCALY PALS

GREEN ANOLE

SEEING A SNAKE slither by or spotting a turtle is a real treat. Reptiles are some of the coolest wildlife around. And not just because they're cold-blooded! Turtles, lizards, snakes, and other reptiles are ancient animals. They've been adapting to life on Earth for hundreds of millions of years. And these dinosaur relatives might be living right in your backyard.

BOX TURTLE

REPTILES ARE MORE COMMON where winters are mild. Lizards don't live in Alaska, for example. But reptiles do thrive in dry regions. Protective scales cover their skin, and reptile eggs have shells that seal in moisture.

AS COLD-BLOODED CREATURES, reptiles soak up heat by sunbathing. If it gets too hot, they simply head for shade or into a cool burrow. Not having to keep their body at a steady temperature saves energy.

SNAKES, LIZARDS, AND TURTLES THRIVE WHERE THEY HAVE PLENTY OF PLACES TO HIDE.

REPTILES NEED LESS FOOD than many birds or mammals. They can go days between meals, which helps them survive in harsh places like deserts. Most reptiles are carnivores who hunt prey.

GARTER SNAKE

WHAT'S THE DIFFERENCE?
LIZARD VS. SALAMANDER

Lizards and salamanders are four-legged, cold-blooded critters with tails. While they are often similar in size, lizards are reptiles and salamanders are amphibians (see chapter 5). Here's how to tell them apart.

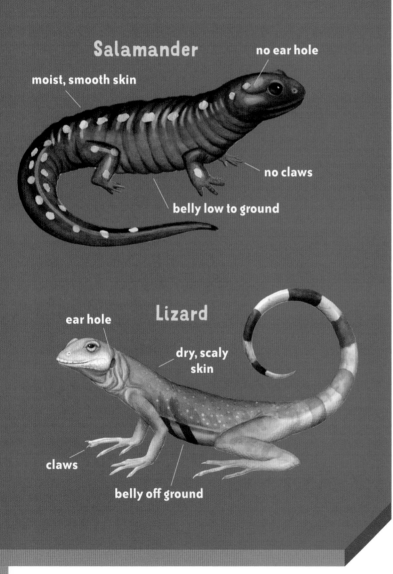

Salamander

moist, smooth skin

no ear hole

no claws

belly low to ground

Lizard

ear hole

dry, scaly skin

claws

belly off ground

Little lizards snack on ants, while big snakes hunt bunnies. Many snakes and lizards get enough water from what they eat.

GOOD REPTILE REFUGES include stone walls, patches of plants, and fallen logs. A heap of rocks is great for sunbathing, and lizards lay eggs in the cracks and crevices. Brush piles provide cover for snakes to hunt and hibernate. Turtles lay eggs in patches of loose sandy soil.

GO NATURAL where you can in the yard. You may be rewarded by the sight of a box turtle munching on a wild strawberry or a garter snake sliding across your path!

BIO BASICS
PARTS OF A
TURTLE

upper shell
(carapace)

tail

eye

nose

mouth

lower shell
(plastron)

front leg

claws

hind leg

MATERIALS

Floor or wall tiles

Spacers such as broken tiles, flat rocks, canning jar lids, tuna cans, etc.

Ruler

Outdoor glue or waterproof caulk

Wooden board (optional)

LIZARD LODGE

Lizards like sunny spots with lots of crevices. Piles of rock or wood work well, but not every backyard has them. Give lizards a place to lounge by building a multistory lodge from leftover floor or wall tiles.

HOW TO

1 There's no set design — just stack up tiles and spacers until you have a lodge you like. The biggest tile makes a good base. If you don't have one big tile, glue together four to six smaller tiles onto a board. Here are some more tips:

- Use spacers to create lizard-friendly heights between levels. Levels can vary and don't have to be perfectly even. They'll be glued.

- Use wide spacers to separate some of the levels into rooms. Lizards like personal space.

- Use the rougher side of the tiles as floors and the slicker side as ceilings.

2 When you have a design you like, glue it together. Starting with the base, put glue on the bottom of the spacers and set them in place on the first layer. Put more glue on top and add the next layer. Keep gluing and stacking until all the pieces are in place.

3 Let the glue dry completely. Place the lizard lodge someplace out of the wind. Choose a spot that gets morning sun, if possible.

MATERIALS

〰〰〰

Shovel or trowel

Plastic, metal, or terra-cotta shallow container (such as a planter saucer, pie tin, or flying disc)

Water

TURTLE SPA

Box turtles are land-loving turtles. They live incredibly long lives, up to a hundred years! In hot weather, box turtles seek out mud to cool off. Make a refreshing mud spa and see if a box turtle comes to visit.

HOW TO

Most reptiles are meat eaters, but box turtles are omnivores who eat a lot of fruit. Here are some native fruits to plant if you can: mayapple, wild black cherry, summer grape, pokeweed, huckleberry, elderberry, native mulberry, blackberry, American persimmon, and crab apple. Use the native plant websites on pages 174–75 to help you choose what grows best where you live.

1 Find a low spot in your yard where water collects when it rains. Dig it out a bit so it stays wet longer.

2 Press the shallow container into the dug-out spot. Add a few handfuls of dirt.

3 Pour in some water to stir up a luxurious mud bath. Check the water level often to keep the spa moist and inviting.

TURTLE PATROL

Many turtles are hit by cars. If you've noticed a spot where turtles cross the road in your neighborhood, post a Watch Out for Turtles sign.

When you're in a car, watch out for turtles on the road and alert the driver when you see one.

If it's safe to help, carry the turtle to the side of the road it's heading toward. Don't put it back where it started because it'll likely try to cross the road again.

WILD NOTES
REPTILES

Wondering what else to take note of?

- Use the Field Guide pages to figure out the snakes, lizards, and turtles you've seen. Keep track of when and where and note what they were doing.

- Draw or photograph the shell patterns of turtles you see. Each is unique, like a fingerprint.

- Make a map of your backyard, noting where reptiles can find cover and habitat.

- Project updates are important! What's working, what's not? Any ideas for improvements?

DON'T PICK UP AFTER YOURSELF!

A messy yard with patches of tall grass and stacked-up brush is great for snakes. Rock piles and left-to-rot logs make good lizard hangouts. Box turtles love leaf litter and patches of bare dirt.

BLACK RAT SNAKE

Pantherophis obsoletus

Large, long snake with squarish, boxy shape because of its flat (not round) belly and straight sides

WHERE DO I LIVE?

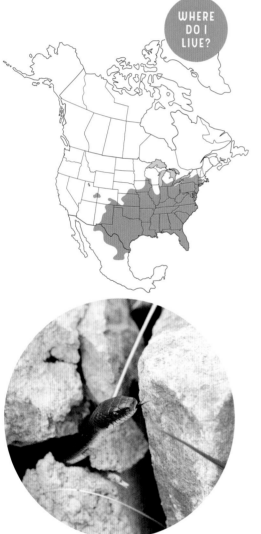

ID ME	SIZE 3 to 7 feet (1 to 2 m) long
	COLOR Black, or black with yellowish stripes, or yellow-gray with stripes on top; throat and belly are light-colored
LOOK FOR ME	Farmlands, leafy forests, swamps, and rocky hillsides
WHAT I EAT	Mice, rats, voles, rabbits, birds, eggs, and lizards
SOME THINGS TO KNOW ABOUT ME	Rat snakes are usually black in the North, but southern rat snakes are yellowish and can have stripes.
	Rat snakes are one of the largest and longest snakes in North America. They're terrific at climbing and easily shimmy up trees and barn poles.
	When picked up by a predator (or person), a rat snake will squirt out a nasty-smelling, musky liquid and smear it around with its tail.

COLLARED LIZARD

Crotaphytus collaris

Big-headed lizard with a long tail and two black, collarlike stripes across the back of the neck

WHERE DO I LIVE?

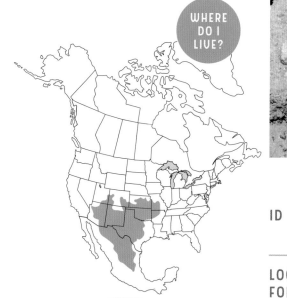

ID ME	SIZE 8 to 14 inches (20 to 36 cm) long
	COLOR Greenish yellow to brown body with light spots; black stripes at back of neck
LOOK FOR ME	Stone ledges and cliffs, rocky hillsides, shrubby canyons, arid woodlands, and dry riverbeds
WHAT I EAT	Insects, spiders, small snakes or lizards
SOME THINGS TO KNOW ABOUT ME	Collared lizards can run on their hind legs like a mini T-rex. And they bite!
	These lizards like to hide under boulders and sunbathe on them, too.
	The color of these western lizards varies a lot. Females are paler but get red or orange spots when carrying eggs. Males are brighter and have bands of color on their backs.

REPTILES

COMMON GARTER SNAKE

Thamnophis sirtalis

Medium-sized snake with three stripes down its back

WHERE DO I LIVE?

ID ME	SIZE 1½ to 3 feet (45 to 91 cm) long
	COLOR Varies a lot! From nearly all black to light-colored with stripes and red checkering
LOOK FOR ME	Neighborhoods, parks, farms, gardens, grasslands, moist woods, and marshes
WHAT I EAT	Worms, frogs, salamanders, and mice
SOME THINGS TO KNOW ABOUT ME	Unlike many snakes, garter snakes don't lay eggs. They give birth to live young.
	The most common snake across the United States and Canada, it's a daytime hunter.
	Garter snakes hibernate in big groups during winter. A den can have hundreds of snakes in it.

EASTERN BOX TURTLE

Terrapene carolina

Land-living turtle with a tall, domed shell that has patches of yellow

WHERE DO I LIVE?

carapace

plastron

ID ME	SIZE 4½ to 6½ inches (11 to 17 cm) long
	COLOR Brown to black top shell with yellow to orange markings
LOOK FOR ME	Woodlands, meadows, old fields, and backyards
WHAT I EAT	Worms, slugs, mushrooms, and fruit
SOME THINGS TO KNOW ABOUT ME	Box turtles live 30 to 40 years in the wild and twice as long in captivity. Males often have red eyes and females have yellowish eyes.
	The ornate box turtle (*Terrapene ornata*) lives in prairies and open woodlands from New Mexico and Texas up through Nebraska and Illinois. Its shell has slash-shaped markings.
	A hinge on the bottom shell (the plastron) allows them to completely close themselves into the top shell (the carapace).

FENCE LIZARD

....

Sceloporus undulatus

....

Small, fast-moving
spiny lizard

WHERE DO I LIVE?

ID ME	**SIZE** 4 to 7½ inches (10 to 19 cm) long
	COLOR Gray to brown; females have barred pattern on back and males have blue throats and markings
LOOK FOR ME	Open forests, brushlands, grasslands, and rocky hillsides
WHAT I EAT	Ants, beetles, and other insects; spiders; centipedes
SOME THINGS TO KNOW ABOUT ME	Each of this lizard's back scales has a backward pointing spine. It's like armor!
	Fence lizards climb trees, walls, and fences to escape danger like a hungry bird, snake, or cat.
	Western fence lizards (*Sceloporus occidentalis*) are bigger and darker. They live west of the Rocky Mountains.

FIVE-LINED SKINK

Plestiodon fasciatus

Smooth, shiny, dark lizard with thin stripes down back and a long tail

WHERE DO I LIVE?

ID ME	SIZE 5 to 8½ inches (13 to 22 cm) long
	COLOR Black with long, yellow stripes; blue to gray tail
LOOK FOR ME	Moist woods, gardens, and forests
WHAT I EAT	Spiders, crickets, caterpillars, beetles, and snails
SOME THINGS TO KNOW ABOUT ME	Rock piles, fallen logs, and other places full of bugs are good places to spot skinks.
	Young five-lined skinks have bright stripes and shiny blue tails. Their stripes fade with age and the tail becomes a dull gray.
	When a predator attacks, the skink drops its blue tail. The tail goes on twitching, keeping the predator interested while the skink escapes. A new tail eventually grows back.

REPTILES

GREEN ANOLE

....

Anolis carolinensis

....

Sleek, greenish lizard with a pointy snout, fat toes, and a throat fan

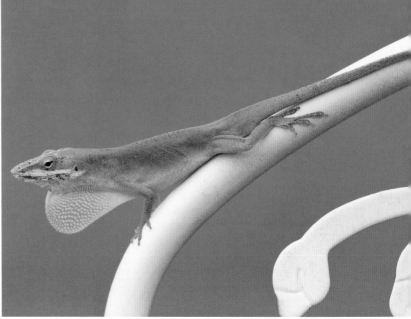

WHERE DO I LIVE?

ID ME	**SIZE** 5 to 9 inches (13 to 22 cm) long
	COLOR Green to brown; males have pink, fan-shaped flap of skin on throat
LOOK FOR ME	Parks, woods, fences, and shrubby areas
WHAT I EAT	Beetles, flies, and other insects; spiders

SOME THINGS TO KNOW ABOUT ME

Male anoles have throat fans. These movable flaps of skin under the chin are like signal flags. They use them to court females and chase off intruders.

Like chameleons (which they aren't), anoles change color to display their moods or camouflage themselves. Being a darker color also lets them soak up more heat on a chilly day.

Brown anoles (*Anolis sagrei*) were introduced from the Caribbean and now compete with native green anoles in Florida and other places along the Gulf Coast.

KINGSNAKE

Lampropeltis getula

Large, shiny, black snake with linking, light-colored bands

WHERE DO I LIVE?

ID ME	SIZE 3 to 5 feet (1 to 1.5 m) long
	COLOR Dark brown to black body with white to yellow bands
LOOK FOR ME	Dry woods, rocky hillsides, prairies, and deserts
WHAT I EAT	Mice, lizards, birds, frogs, and snakes
SOME THINGS TO KNOW ABOUT ME	Kingsnakes hang out in log and brush piles. They hunt at dawn and dusk, or at night during hot weather.
	Kingsnakes hunt rattlesnakes and copperheads and are immune to their venom.
	Kingsnakes and their milk snake cousins are constrictors. They kill by wrapping themselves around prey and squeezing.

REPTILES

PAINTED TURTLE

Chrysemys picta

Medium-sized, freshwater turtle with red markings on the edges of its shell and on its neck

WHERE DO I LIVE?

ID ME	SIZE 4 to 9 inches (10 to 23 cm) long
	COLOR Olive to black top shell with red marks on edges; orange bottom shell; black-and-yellow striped head; red lines on neck
LOOK FOR ME	Marshes, ponds, shallow lakes, slow creeks, and rivers
WHAT I EAT	Water plants and animals, including insects and fish
SOME THINGS TO KNOW ABOUT ME	Painted turtles warm up by basking on logs. If there isn't much space, they'll stack up on top of each other.
	Young turtles need meat to grow properly, but older turtles eat mostly plants.
	The painted turtle is the most widespread native turtle in all of North America. Fifteen-million-year-old fossils of it have been found.

RINGNECK SNAKE

························

Diadophis punctatus

························

Small, slender snake with a bright belly and neck ring

WHERE DO I LIVE?

ID ME	SIZE 10 to 30 inches (25 to 75 cm) long COLOR Gray-green to black on back, yellow to red belly, orange ring around neck
LOOK FOR ME	Moist forests, grasslands, and along desert streams
WHAT I EAT	Worms, slugs, salamanders, and small lizards
SOME THINGS TO KNOW ABOUT ME	This secretive, shy snake hides and hunts under flat stones and leaf litter. Ringneck snakes are harmless but let loose stinky poop and pee when picked up. When some ringnecks feel threatened, they wrap into a coil and stick their tail up in the air while protecting their head. Better a bite to the tail than to the head!

REPTILES

GRAY TREEFROG

·5·

WELCOME

FROGS, SALAMANDERS

& OTHER SLIPPERY NEIGHBORS

RED EFT

THERE'S SOMETHING WONDERFUL ABOUT FINDING A NEWT under a lifted rock or hearing frogs calling at night. Hosting slippery creatures like these in your yard means you're doing something right. They can't live just anywhere.

FROGS, TOADS, NEWTS, AND SALAMANDERS are amphibians. Like reptiles, they are cold-blooded and more common in milder regions. But most amphibians hatch from eggs laid in water and live at first as a water creature.

TADPOLES AND BABY SALAMANDERS breathe with gills and swim like fish. They spend weeks or months in water before sprouting legs and starting their lives on land. Living between two worlds makes amphibians vulnerable to pesticides and pollution both on land and in water.

NORTH AMERICAN FROGS, TOADS, NEWTS, AND SALAMANDERS are meat-eating predators. Adults hunt whatever bug, fish, slug, or other animal they can catch and swallow. Tadpoles and baby salamanders are less picky. Most eat whatever is in the water, including algae and plants.

FROG

YARDS WITH TOADS, SALAMANDERS, AND THEIR KIN ARE A SIGN OF A HEALTHY HABITAT.

BULLFROG

SALAMANDER

119

WHAT'S THE DIFFERENCE?

Frogs and toads are amphibians. They are more alike than different. Here's how to tell them apart.

Toad

thick, dry, bumpy skin

round, plump body

shorter legs that walk or hop

Frog

thin, wet, smooth skin

longer, slim body

longer legs that leap

EVEN LAND-LIVING AMPHIBIANS need to stay moist to breathe. If their smooth, slimy skin dries out, they suffocate. Frogs, newts, and other amphibians need shelter from the sun and weather. Damp, rotting logs, shady, dense plants, and rocks in moist low places are amphibian havens.

SALAMANDERS, NEWTS, FROGS, and toads don't build nests. Most lay eggs in watery spots — ponds, rain-filled ditches, or slow streams. Springtime pools of water that dry up in summer are especially great for frogs and salamanders. There are no fish in them to eat the tadpoles and baby salamanders. Letting a swampy area of your yard simply be is a great help to these backyard wonders.

MATERIALS

Hard plastic
wading pool
or other large
shallow tub

Sand or gravel

Shovel

Rocks

Water

Water plants
(optional)

KIDDIE-POOL FROG POND

Ponds are wildlife magnets that provide all of the Big Four habitat requirements! A permanent pool of water provides shelter and nesting places for aquatic bugs, a drinking spot for all sorts of animals, and a must-have habitat for frogs. A pond is where frogs find mates, breed, and lay their jellylike eggs.

HOW TO

1 Find a low, shady, sheltered area at the bottom of a hill or slope. That way rain will help keep the pool full. Turn the pool upside down in the spot you've chosen and use some sand or gravel to create an outline. Remove the pool.

2 Dig out the area you outlined. Pile up the dirt in one spot; you'll need it later. Set the pool into the hole, digging out or filling in with dirt to get it to fit. The pool doesn't have to be level, but the bottom has to be supported or it will crack over time.

3 Arrange some flat rocks all around the outside edge of the pond, using leftover dirt to level and secure them. Place a couple of large rocks or bricks in the pool, especially near the inside edges. These help the frogs get in and out.

4 Dump an inch (2.5 cm) or so of sand or gravel into the pool. Cover the bottom of the pool with several shovelfuls of dirt. Add a layer of gravel or sand. Fill the pond with water and add a few aquatic plants if you want. A consistent water level is best, so add more water as needed.

TREE FROG HIDEOUTS

Want to help tree frogs? These homemade hideouts
are perfect perches for their nighttime bug hunts.
During the day, the hideouts provide safe, shady
shelter from predators.

HOW TO

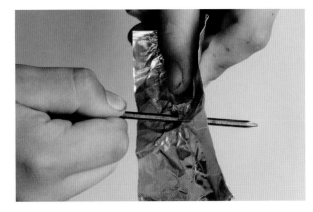

1 Cut a 3-inch (7.5 cm) strip of foil duct tape. Fold it in half lengthwise, sticky sides together. Fold over any sticky surfaces along the edges. Use the nail to poke a hole in the center of the folded tape.

2 Place the folded duct tape over one end of the PVC pipe, centering the hole so rainwater will drain. Attach it with more duct tape wrapped around the pipe.

3 Find a shady tree, near water if possible. Strap your tree frog hideout onto the tree's trunk with bungee cords. Put the open end at eye level so you can see who's in there!

• NOTE •

If you live near the tropics, like in southern Florida, this version is better suited to your climate:

1. Use a 3-foot (91 cm) length of PVC pipe and ask an adult to saw off one end at a slant.

2. Push the slanted end firmly into the ground near some lush plants.

3. Drill or hammer a drain hole a few inches (5–6 cm) above the soil level.

MATERIALS

Outdoor glue or waterproof caulk

Terra-cotta or plastic plant pots, broken or not

Paint and/or decorations (optional)

Hand trowel

Leaf litter and moss (optional)

TOAD ABODES

Toads spend more time on land than frogs do. These bumpy-skinned amphibians are prey for snakes and birds so are always looking for places to hide. Help them out by adding some toad-friendly hangouts to cool, damp, shady spots in your yard. Gardeners will thank you, too! Toads eat lots of slugs and other veggie-gobbling pests.

HOW TO

1 Glue broken pots or terra-cotta pieces together to make a toad house with a doorlike opening. Or use an unbroken pot.

2 Decorate the toad abode, if you'd like. Paint it or glue on glass globs or shells. If your pot is unbroken, only decorate half of the pot. (The other half will be buried in the ground.)

3 Let your pot dry.

4 Choose a damp shady place. Bury a whole pot halfway to make a cave. Sink a glued-up upright pot into the soil an inch (2.5 cm) or so. Some leaf litter inside and moss outside make it toad-ally welcoming.

You can hear the calls of different frogs on the websites listed on page 174.

WHO'S THAT CALLING?

Every species of frog has a different call. Learning frog calls lets you know who's around.

128

TIPS FOR ATTRACTING FROGS TO A POND

- Don't install a fountain. Frogs like still water.

- Forget the fish; they eat frog eggs.

- Add some pond plants like lily pads.

- Don't let the water level get too low, especially in summer.

- Provide hiding places and cover around the pond. Let grass or weeds grow high, or create piles of leaves, logs, rocks, or brush.

WILD NOTES
AMPHIBIANS

Wondering what else to take note of?

- Use the Field Guide pages to figure out which frogs, toads, and salamanders you've seen. Keep a list of all the amphibians, when you saw them, and where. Include egg masses and tadpoles.

- Project updates are important! What's working, what's not? Any ideas for improvements?

- Map a toad's-eye-view route through the yard and fill in the gaps with logs or rocks that provide cover.

EMBRACE THE MESS!

Frogs and toads seek out the dark, damp spaces under piles of leaves, brush, and logs. Patches of tall grass and thick plants provide protection from the sun and predators.

AMERICAN BULLFROG

....

Lithobates catesbeianus

....

Very large, green frog with big, bulgy, gold eyes and wide mouth

WHERE DO I LIVE?

ID ME	SIZE 4 to 8 inches (10 to 20 cm) long
	COLOR Green or green with dark splotches on top, light belly
	VOICE Loud *jug-O-rum, jug-O-rum*
LOOK FOR ME	Ponds, lakes, or slow streams with plants
WHAT I EAT	Bugs, fish, frogs, snakes, and birds
SOME THINGS TO KNOW ABOUT ME	The largest frog in North America, a bullfrog can weigh more than a pound (0.5 kg).
	Bullfrogs will eat whatever they can catch, which is a lot because they're so big. Baby ducks, songbirds getting a drink, and anything in the water is fair game.
	Bullfrogs become pests when they're introduced into non-native habitats. They'll eat everything else that lives in a pond.

AMERICAN TOAD

Bufo/Anaxyrus americanus

Bumpy, plump, short-legged hopping toad with a light-colored line down its back and a round snout

WHERE DO I LIVE?

ID ME	SIZE 2 to 4 inches (5 to 10 cm) long
	COLOR Brown, gray, and/or rusty red with bumpy spots
	VOICE Long, cricketlike trill
LOOK FOR ME	Gardens, woods, ponds, and lakes
WHAT I EAT	Insects, worms, slugs, and snails
SOME THINGS TO KNOW ABOUT ME	Woodhouse's toad (*Bufo woodhousii*) has a pointier snout and darker brown blotchy back. It's more common west of the Mississippi River.
	Toads live in drier areas than most frogs but need water to breed and lay eggs.
	Unlike most toads, American toads have long, sticky, bug-zapping tongues.

AMPHIBIANS

GRAY TREE FROG

Hyla versicolor

Frog with long legs, round suction-cup toes, bumpy skin on back, and dark eyes

WHERE DO I LIVE?

ID ME	SIZE 1 to 2 inches (3 to 5 cm) long
	COLOR Brown to green to gray with dark blotches on back and yellow-orange inside thighs
	VOICE Medium chirplike trill
LOOK FOR ME	Shrubs and trees near water
WHAT I EAT	Insects
SOME THINGS TO KNOW ABOUT ME	This nocturnal hunter comes out at night to find tasty bugs.
	The gray tree frog rarely comes down to the ground, spending most of its time in trees and bushes.
	Gray tree frogs prefer to breed on branches hanging over a pond or shallow water. That way the eggs fall right in! Tadpoles hatch out of the eggs after about a week, sooner if the water is warm.

GREEN FROG

Lithobates clamitans

Green to brown frog with back ridges and bulgy eyes

WHERE DO I LIVE?

ID ME	SIZE 2 to 4 inches (5 to 10 cm) long
	COLOR Green to brown
	VOICE One-note, loose-banjo-string
	SOUND *twaaanng*
LOOK FOR ME	Shallow streams, ponds, and springs
WHAT I EAT	Bugs, frogs, and small fish
SOME THINGS TO KNOW ABOUT ME	Big green frogs can look like small bullfrogs. Check for back ridges, which bullfrogs don't have.
	Green frogs are more brownish bronze in color in the South and greener in the North.
	Males have eardrums twice as big as their eyes; female eardrums are smaller.

AMPHIBIANS

RED-BACKED SALAMANDER

Plethodon cinereus

Long, skinny, dark salamander with orangish back stripe

WHERE DO I LIVE?

ID ME	**SIZE** 2½ to 4 inches (6 to 10 cm) long
	COLOR Dark gray to black body with small light spots; some have yellow to red stripe on back
LOOK FOR ME	Under logs, rocks, or moist leaves in woods
WHAT I EAT	Mites, spiders, millipedes, snails, ants, beetles, and earthworms
SOME THINGS TO KNOW ABOUT ME	Like many salamanders, they can drop off their tail to escape a predator. A new one eventually grows back.
	Unlike most amphibians, red-backed salamanders don't have a water life stage. Tiny salamanders hatch out of eggs laid under a log or in a burrow.
	Red-backed salamanders are important insect predators that eat garden pests.

RED-SPOTTED NEWT

Notophthalmus viridescens

Slender, bright-red, rough-skinned salamander with spots, found under leaves or logs

WHERE DO I LIVE?

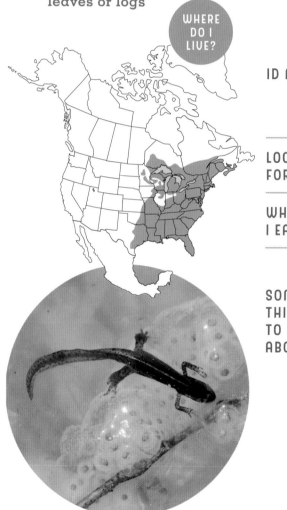

ID ME	**SIZE** 2½ to 5 inches (6 to 12 cm) long **COLOR** Adult: Olive on top with red spots outlined in black; yellow below with small black spots. Eft (young newt): Bright reddish-orange all over with red spots outlined in black on back
LOOK FOR ME	Adults in ponds and swamps; efts in moist woods
WHAT I EAT	Snails and insects on land; snails, insects, worms, fish eggs, and tadpoles in water
SOME THINGS TO KNOW ABOUT ME	These newts spend two to seven years as red efts, living on land. Once ready to breed, efts change into adults and move to water. Water-living adults stay active all year. In winter they sometimes can be seen swimming under ice. Many predators leave red-spotted newts alone because their skin oozes an irritating and bad-tasting toxin.

AMPHIBIANS

SPRING PEEPER

Pseudacris crucifer

Small, brownish frog with a darker X marking on back

WHERE DO I LIVE?

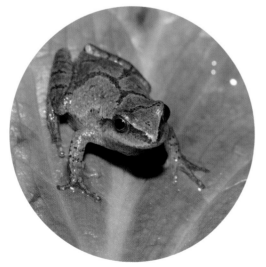

ID ME	SIZE ½ to 1 inch (1 to 3 cm) long
	COLOR Tan to brown to gray
	VOICE High-pitched, whistlelike *preep, preep*
LOOK FOR ME	Wooded areas with ponds, swamps, or temporary spring pools
WHAT I EAT	Insects, spiders, and small bugs
SOME THINGS TO KNOW ABOUT ME	These small frogs spend winters hibernating under logs or tree bark. In very early spring, the males gather in groups and call to the silent female frogs.
	Spring peepers are a kind of chorus frog. All are early spring singers. Depending on where you live, there are western chorus frogs (*Pseudacris triseriata*) and spotted chorus frogs (*Pseudacris clarkii*).
	One small spring peeper female will lay as many as a thousand eggs between March and June.

TIGER SALAMANDER

Ambystoma tigrinum

Large, plump, big-headed, and dark-colored salamander covered in yellowish markings

WHERE DO I LIVE?

ID ME	SIZE 6 to 10 inches (15 to 25 cm) long
	COLOR Brown to black body with yellow to olive spots and splotches
LOOK FOR ME	Woodlands, grasslands, and forests near water
WHAT I EAT	Worms, snails, insects, and slugs
SOME THINGS TO KNOW ABOUT ME	The tiger salamander is the largest land-living salamander in North America.
	These big salamanders dig burrows up to 2 feet (61 cm) deep. They spend most of the year underground.
	In general, salamanders eat whatever they can catch and fit into their mouth. Depending on their size, that might include frogs, baby snakes, newborn mice, and even smaller salamanders.

EASTERN COTTONTAIL

·6·

FOX

HELP OUT YOUR
FURRY
FRIENDS

SKUNK

IMAGINE SEEING A FOX slipping under a fence, a deer drinking from a pond, or a woodchuck nibbling clover in the lawn. Spotting furry wild animals in your backyard is something special. It's like you live in the wilderness, too!

ANIMALS WITH FUR OR HAIR are mammals. They give birth to live young that drink mother's milk. (Yep, you're a mammal!) Like birds, mammals are warm-blooded. They need to eat a lot of food to fuel their steady body temperature. Some are meat-eating predators, while others munch berries, nuts, seeds, and plants.

MANY MAMMALS ARE PREY for owls and hawks, snakes and coyotes. Voles, rabbits, and squirrels are food for other backyard wildlife. Your yard is part of the web of life!

MOST MICE AND OTHER SMALL MAMMALS get water from eating juicy plants or drinking dew. But larger mammals will travel miles to ponds, birdbaths, and other water sources — especially if you live someplace with a dry or hot season.

NORTH AMERICAN MAMMALS RANGE FROM SMALL BATS WEIGHING LESS THAN A POTATO CHIP TO GRIZZLY BEARS STANDING 10 FEET (3 M) TALL.

CHIPMUNK

SKUNK

139

DEER

WHITE-TAILED DEER

A GOOD MAMMAL HABITAT also includes shelter. Bushes, log piles, tall grass, and rock stacks are places for chipmunks to hide, foxes to hunt, and rabbits to make nests. Squirrels depend on large trees for nuts and nesting spots, and hollow logs are favorite sleeping spots for raccoons, foxes, and opossums.

IN COLD CLIMATES, MAMMALS can have a hard time finding food in winter. Some insect-eating bats simply fly south. Other mammals save energy by sleeping away the winter underground or by snacking on stored food like dried grasses or seeds.

TRUE HIBERNATING MAMMALS, like cave bats and woodchucks, don't eat or drink. Their bodies slow down and survive on stored fat. Many mammals need to eat a year's worth of food during the warmer months. Another reason to make your backyard into a mammal-friendly habitat!

tail

BIO BASICS
PARTS OF A
GRAY
SQUIRREL

ears

back

eye

nose

mouth

whiskers

front leg

hind leg

belly

claws

MATERIALS

Outdoor glue or waterproof caulk

Paint-stirring stick

Large wide-mouthed jar or tub

Nuts or birdseed

Wire or bungee cord (optional)

EAT-IN DINER

Invite some chipmunks or squirrels in for a snack with this see-through feeder. You'll be able to get a good look at the action!

HOW TO

1 Glue the stick onto one side of the jar.

2 Fill the bottom of the jar with nuts or birdseed.

3 Choose an easy-to-watch place for your diner. You can attach the feeder to a deck railing, post, or low branch by wrapping wire or a bungee cord around the stick.

Or if you know chipmunks are living in a stone wall or a tree, just set the diner on the ground near their home. Get your camera ready!

NUTTY WREATH

Squirrels love peanuts! Make a fun wreath full of this nutty treat from a spring toy and wire hoop. Many birds — like wrens, nuthatches, and blue jays — will thank you, too.

HOW TO

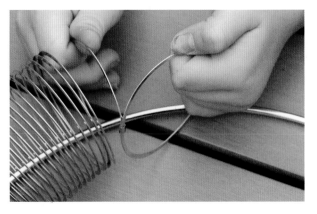

1 If the top and bottom coil of your spring toy are closed loops, either pry off the band to open the loop or bend that first coil out of the way.

• NOTE •

If you're using a hoop made from a hanger, just thread the spring toy on it, twist the ends of the hanger together with pliers, and go to step 3.

2 Hold one end of the spring toy. Wrap the first coil around the hoop. Keep going until the whole spring toy is on the hoop. It takes a while!

3 Use paper clips to join the ends of the spring toy. Lay the hoop flat and spread the coils evenly around it.

4 Zip tie a coil to the hoop on the opposite side from the paper clips. Make it tight! Add another zip tie every five or so coils. The spring shouldn't sag. Cut off the zip-tie tails.

5 Fill the wreath with unshelled peanuts. Attach a string or cord to hang the nutty wreath from a tree or hook.

Want to build your own bat house? Look for the web address on page 175.

PAINT A BAT HOUSE

Bat houses attract these mosquito-munching flying mammals by giving them places to roost and raise young. But before putting one up, make sure it's a suitable home.

The color of the house matters. In colder regions, use dark colors that will soak up heat from the sun.

Many store-bought bat houses are too small. It needs to be *at least* 2 feet (61 cm) tall and 14 inches (36 cm) wide.

Bats need a landing area below the entrance or inside.

MIDNIGHT GARDEN

Bats eat insects, not plants. But you can plant flowers that attract tasty nighttime bugs. Try fleabane daisies, evening primrose, phlox, sleepy catchfly, and goldenrod. Use the native plant websites on pages 174–75 to help you choose what grows best where you live.

PHLOX

EVENING PRIMROSE

WHAT'S THE DIFFERENCE?

When the sun sets, bats come out and birds head to bed. In the low light of evening, it can be hard to tell bird from bat. Here's what to look for.

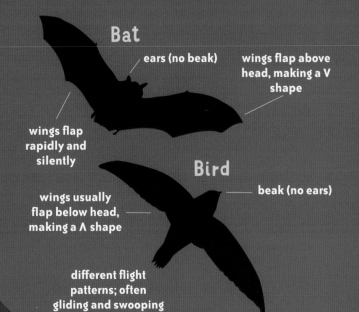

Bat

ears (no beak)

wings flap above head, making a V shape

wings flap rapidly and silently

Bird

beak (no ears)

wings usually flap below head, making a ∧ shape

different flight patterns; often gliding and swooping

WHAT COLOR SHOULD YOUR BAT HOUSE BE?

Find where you live on the map. Use the recommended color, using a water-based stain or flat outdoor paint.

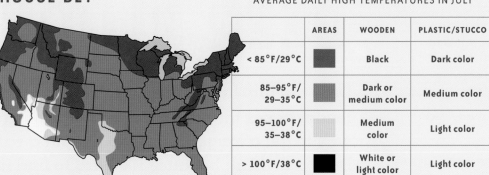

BAT HOUSE COLOR RECOMMENDATIONS AND AVERAGE DAILY HIGH TEMPERATURES IN JULY

	AREAS	WOODEN	PLASTIC/STUCCO
< 85°F/29°C		Black	Dark color
85–95°F/ 29–35°C		Dark or medium color	Medium color
95–100°F/ 35–38°C		Medium color	Light color
> 100°F/38°C		White or light color	Light color

MATERIALS

½ cup (120 mL) vegetable oil

1 cup (240 mL) brown sugar

1 cup (240 mL) molasses

2 tablespoons (35 mL) salt

Large mixing bowl

Mixing spoon or spatula

2 cups (480 mL) cornmeal

2 cups (480 mL) old-fashioned oats

Roasting pan

Cooking spray

2-cup (480 mL) disposable plastic container

Plastic drinking straw

Cord or string

Chopstick or skewer

Short, sturdy stick

SWEET & SALTY DEER BLOCK

Deer are big, hungry critters! They can empty a bird feeder or eat through a flower garden in a flash. One way to get a longer look is to put out a salty food block. Deer crave salt because it's hard to find in the wild. Get ready to watch tongues in action.

HOW TO

1 Preheat the oven to 325°F (170°C). Put the oil, sugar, molasses, and salt into a large bowl. Stir. Then mix in the cornmeal and oats. Stir until well blended.

2 Cover the bottom of the roasting pan with cooking spray. Empty the oats mixture onto the pan. Bake for 15 minutes.

Stir the mixture and return to the oven for 15 minutes. Remove from the oven and let cool for 10 minutes.

3 Coat the plastic tub with cooking spray. Spoon about 2 inches (5 cm) of still-warm mixture into the tub and press down to compact it.

4 Coat a straw with cooking spray. Push the straw into the center of the mixture, all the way to the bottom of the tub. Spoon the rest of the mixture into the tub, pressing it down as you go. Let it set up overnight at room temperature.

5 Pop out the block and cut the straw flush with the top. Cut a length of cord about 2 feet (0.6 m) long. Double it and thread the bent end through the straw, pushing it through with the chopstick. Slide the stick into the loop at the bottom. Tie the loose ends into a tight knot to keep the stick in place. Hang the block where deer can reach it, a few feet off the ground.

CORN COB JUNGLE GYM

Twist screw eyes or use a screwdriver to insert screws into the tops of several dried corncobs.

Tie a piece of sturdy elastic to each one (or use thin bungee cords). Add a dab of glue in the screw holes if the screws aren't holding; let dry.

If squirrels are scaring away your feathered friends from the feeders, give them something else to eat.

Hang the bouncy corncobs from a branch at different heights and let the squirrel acrobatics begin!

MAKE SPACE FOR MAMMALS

Turn yard waste into some mammal-friendly space! Moles appreciate leaf piles, while shrews like stacks of rock. Brush piles are great for rabbits, chipmunks, woodchucks, and other creatures looking for shelter, cover from predators, and places to nest.

WEBCAM WILDLIFE WATCH

- Turn a webcam toward a window bird feeder and see who's snacking at seeds while you're away.

- If you're wondering what larger wildlife might be wandering around your yard, consider using a trail camera. These inexpensive all-weather devices automatically snap a photo when something walks in front of it.

WILD NOTES
mammals

- Use the Field Guide pages to figure out which furry creatures you're seeing. Keep a list of all the mammals spotted, when you saw them, and where.

- Got a trail camera? Map out your yard, keeping track of where you place the camera. Note its height and lens direction, too. Print out any photos and include them with notes.

- Sketch any tracks you see and try to identify them.

- Project updates are important! What's working, what's not? Any ideas for improvements?

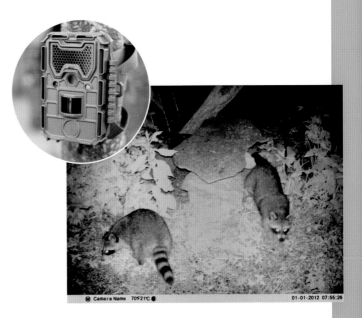

Camera Name   70°F21°C   01-01-2012 07:55:26

BIG BROWN BAT

Eptesicus fuscus

Medium-sized, fast-flapping, furry brown bat with darker ears and face

WHERE DO I LIVE?

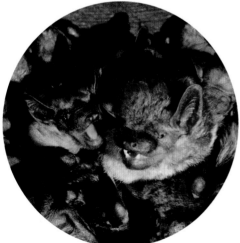

ID ME	BODY About 3 inches (7.5 cm) long
	WINGSPAN 13 inches (33 cm)
	SIGNS Smelly, dark, grain-shaped poop under roosts
LOOK FOR ME	Forests, farmlands, parks, and neighborhoods
WHAT I EAT	Beetles, mosquitoes, and moths
SOME THINGS TO KNOW ABOUT ME	Big brown bats sleep away the day in hollow trees and old barns. In winter they hibernate in caves, old mines, attics, or abandoned buildings. They are hardy and put up with people living nearby.
	Big brown bats are fast fliers and reach speeds up to 40 miles (64 km) per hour.
	Bat poop, or guano, collects in giant piles under their roosts. It stinks, but gardeners love it because it enriches the soil!

BLACK BEAR

Ursus americanus

Dark bear with paler muzzle and round back

WHERE DO I LIVE?

ID ME	**BODY** 4 to 6 feet (1.2 to 1.8 m) long **SIGNS** Ripped-apart rotten logs; claw marks on trees; large 6- to 7-inch (15 to 18 cm) tracks with five toes; scat like big dog droppings with seeds and insect shells in it
LOOK FOR ME	Forests, swamps, parks; suburbs in the East
WHAT I EAT	Fruit, berries, nuts, seeds, plants, insects, carrion, and some small prey
SOME THINGS TO KNOW ABOUT ME	Black bears aren't always black. Some are even blond! West of the Rocky Mountains, black bears are often dark brown or red-brown. There's a big size difference between males and females. Males average 265 pounds (120 kg) and females 175 pounds (80 kg). Black bear populations are increasing across North America. They're even showing up at backyard bird feeders and in neighborhood parks.

MAMMALS

BLACK-TAILED PRAIRIE DOG

Cynomys ludovicianus

Big, tan ground squirrel with white belly and black-tipped tail

WHERE DO I LIVE?

ID ME	BODY About 10 inches (25 cm) long
	SIGNS Bunch of burrows with hard-packed dirt domes around each hole; *wee-oo* barks between neighbors
LOOK FOR ME	Short-grass prairies
WHAT I EAT	Grasses

SOME THINGS TO KNOW ABOUT ME	Prairie dogs often stand on their hind legs to get a good view of who is around. When they see or hear danger, they sound the alarm and everyone heads underground.
	Groups of prairie dogs live in large colonies called towns. Town members cooperate to look out for predators, care for young, and collect hay for winter.
	Prairie dogs keep their town's grass clipped short so the view is clear and hungry snakes can't sneak in.

COYOTE

....

Canis latrans

....

Gray- to rusty-tan-colored medium wild dog with pointy ears and hanging bushy tail

WHERE DO I LIVE?

Western coyote

Eastern coyote

ID ME	BODY About 3 feet (1 m) long; tail another 12 to 15 inches (30 to 38 cm)
	SIGNS Doglike scat full of hair; yipping sound; doglike tracks about 2.5 inches (6 cm) long
LOOK FOR ME	Open plains, forest edges, parks, and suburbs
WHAT I EAT	Rabbits, mice, and other small- to medium-sized prey, carrion, fruit; and garbage
SOME THINGS TO KNOW ABOUT ME	Coyotes were once a western critter. But with wolves now gone from most of the continent, they've found new territory and prey.
	Eastern coyotes are bigger and hunt deer. Interbreeding with wolves (*Canis lupus*) near the Great Lakes while moving east may have supersized them.
	Coyotes are smart and flexible, adapting and learning to survive in all kinds of places. Some are loners, while other coyotes live in packs or with a mate.

MAMMALS

DEER MOUSE

Peromyscus maniculatus

Mouse with brown back, white belly, long white-bottomed tail, and large shiny black eyes

WHERE DO I LIVE?

ID ME	SIZE Body 3 to 4 inches (8 to 10 cm) long; tail another 2 to 5 inches (5 to 12 cm)
	SIGNS Tiny tracks with four toes and claw marks; nests in trees, burrows, or hollow logs
LOOK FOR ME	Woodlands, fields, grasslands, suburbs, and farmland
WHAT I EAT	Seeds, nuts, fruits, insects, and fungi
SOME THINGS TO KNOW ABOUT ME	Deer mice are good climbers and often live in trees and old bird nests.
	A similar cousin is the white-footed mouse (*Peromyscus leucopus*). But its tail is gray, not brown on top and white below like the deer mouse.
	Don't confuse them with house mice (*Mus musculus*) which have gray bellies, naked tails, and are non-native pests.

EASTERN CHIPMUNK

Tamias striatus

Small, red-brown animal with white and dark stripes along sides and long furry tail

WHERE DO I LIVE?

ID ME	**SIZE** Body about 6 inches (15 cm) long; tail another 3 to 4 inches (8 to 10 cm)
	SIGNS Chewed nuts on logs and rocks; *chip-chip-chip* sound; small burrows with openings hidden within tree roots or under rock walls
LOOK FOR ME	Leafy forests, suburbs, brushy areas, and stone walls
WHAT I EAT	Seeds, nuts, berries, and fungi
PLANT THIS FOR ME	Oak, hickory, beech, and walnut trees
SOME THINGS TO KNOW ABOUT ME	Chipmunks dig burrow systems with lots of holes, chambers, and escape tunnels. Places with stumps, rocks, and fallen logs are chipmunk friendly. Chipmunks need places to perch near hidden burrow holes. Where winter is harsh, chipmunks hibernate. In milder places, or during warmer years, they wake up throughout the winter to snack on food they stored in fall.

MAMMALS

EASTERN COTTONTAIL

Sylvilagus floridanus

Gray-brown rabbit with a fluffy white tail, white feet, and rusty neck fur

WHERE DO I LIVE?

ID ME	**BODY** 13 to 16 inches (33 to 40 cm) long
	SIGNS Small, 1-inch (3 cm), round front tracks with hind tracks three times as long; scat looks like small, round, dark balls
LOOK FOR ME	Old farm fields, suburbs, gardens, parks, swamps, prairies, and forests
WHAT I EAT	Grasses and other plants in summer; bark, twigs, and buds in winter
PLANT THIS FOR ME	Native clovers, grasses, and bushes
SOME THINGS TO KNOW ABOUT ME	One female cottontail can have six or seven litters of five or more kits in a year. That's a lot of bunnies!
	Rabbits often feed at night. But you can also see them at dawn and dusk.
	Lots of animals prey on cottontails. They run in a fast zigzag dash when escaping a predator.
	The black-tailed jackrabbit (*Lepus californicus*) is a larger, longer-eared cousin of the cottontail that lives in the western United States and Mexico.

EASTERN GRAY SQUIRREL

....

Sciurus carolinensis

....

Gray to gray-brown, big tree squirrel with long bushy tail

WHERE DO I LIVE?

ID ME	SIZE Body about 9 inches (23 cm) long; tail the same length
	SIGNS Chewed nuts on ground; leaf nests in trees; front tracks are 1.6 inches (4.1 cm) long with four toe prints while hind tracks are 2.6 inches (6.6 cm) with five toes
LOOK FOR ME	Leafy forests with nut trees, neighborhoods, and cities
WHAT I EAT	Acorns and other nuts, seeds, buds, bark, fungi, and flowers
PLANT THIS FOR ME	Oak, hickory, beech, and walnut trees
SOME THINGS TO KNOW ABOUT ME	Squirrels are important spreaders of tree seeds. Not all the nuts they bury as winter food are eaten. Some sprout and grow into trees that provide more seeds.
	Their nests, called dreys, look like big clumps of leaves stuffed into crooks of branches high in trees.
	Reddish fox squirrels (*Sciurus niger*) like more open areas and spend more time on the ground than smaller gray squirrels.

MAMMALS

MEADOW VOLE

Microtus pennsylvanicus

Furry, round, chubby brown relative of the mouse with little legs, small ears, and a short tail

WHERE DO I LIVE?

VOLE PATHS

ID ME	**SIZE** Body 4 to 5 inches (10 to 13 cm) long; tail another 1 to 2 inches (2.5 to 5 cm)
	SIGNS Networks of narrow pathways through grasses and plant-covered fields that lead to burrow holes; tracks are 5 inches (13 cm) long with five toes on hind print and four toes on front print
LOOK FOR ME	Fields, meadows, highway medians, and marshes
WHAT I EAT	Plants in summer and seeds, bark, and roots in winter
SOME THINGS TO KNOW ABOUT ME	The meadow vole is also called meadow mouse or field mouse. They love tunneling in fields of hay and other crops and are considered a pest by many farmers and gardeners.
	Meadow voles are food for everyone from fish to bears. But they breed quickly and year-round so are often plentiful. A single acre can support many hundreds of voles.
	The woodland vole (*Microtus pinetorum*) looks similar but lives in forests east of the Mississippi River.

MEXICAN FREE-TAILED BAT

Tadarida brasiliensis

Medium-sized brown bat with large ears and a pointy tail that's not attached to a flap of skin

WHERE DO I LIVE?

ID ME	SIZE Body 2 to 3 inches (5 to 8 cm) long; tail another 1.5 inches (4 cm)
	WINGSPAN 11 inches (28 cm)
	SIGNS Smelly, brown, rice-shaped droppings under bridges or in caves
LOOK FOR ME	At dusk in farmlands, deserts, and open areas
WHAT I EAT	Moths, beetles, and other flying insects
SOME THINGS TO KNOW ABOUT ME	Mexican free-tailed bats spend winters in central and southern Mexico. They are also called Brazilian free-tailed bats.
	These bats are important controllers of farm pests, especially moths that lay eggs in corn and cotton.
	Tens of millions of Mexican free-tailed bats spend their summers in the southern half of the United States. Groups of thousands of mother bats and their pups snuggle into caves and squeeze into cracks under bridges to sleep during the day.

MAMMALS

MULE DEER

Odocoileus hemionus

Beefy deer with large "mule" ears and black-tipped tail

WHERE DO I LIVE?

ID ME	**BODY** About 6 feet (2 m) long
	SIGNS Tracks from narrow split hooves; dark oval scat pellets almost 1 inch (2 cm) long; small trees with missing bark from males rubbing antlers; patches of flattened grass or snow from napping deer
LOOK FOR ME	Forests, prairies, and mountains
WHAT I EAT	Sprouting and young grasses, twigs, and shrubs
SOME THINGS TO KNOW ABOUT ME	Male mule deer, called *bucks*, shed antlers in January and regrow them over the summer.
	In mountain areas, mule deer migrate down to the foothills in winter to find food.
	Female mule deer give birth to one or two spotted fawns.

NORTHERN SHORT-TAILED SHREW

·······························

Blarina brevicauda

·······························

Mouse-sized, gray to brown animal with velvety fur, a long snout, no visible ears, tiny eyes, and a short tail; makes buzzing sounds

WHERE DO I LIVE?

ID ME	SIZE Body 3 to 4 inches (8 to 10 cm) long; tail another inch (2.5 cm)
	SIGNS Grass or leaf nests under a log
LOOK FOR ME	Forests, grasslands, and suburbs
WHAT I EAT	Insects, seeds, and worms
SOME THINGS TO KNOW ABOUT ME	Shrews aren't gnawing rodents, like mice. They're insectivores, like moles.
	Northern short-tailed shrews taste so bad that dead ones are often left uneaten.
	Short-tailed shrews spend most of their time out of sight, tunneling under dead leaves looking for food.

MAMMALS

PORCUPINE

Erethizon dorsatum

Big, round, slow-moving animal covered in long, thick spines called quills

WHERE DO I LIVE?

ID ME	**SIZE** Body about 2 feet (61 cm) long; tail another 6 to 12 inches (15 to 30 cm)
	SIGNS Chewed tree bark and nipped twigs; piles of oval droppings at the bottoms of trees
LOOK FOR ME	Rocky pine forests in the East and dry, brushy areas in the West
WHAT I EAT	Tree bark in winter and plants in summer
SOME THINGS TO KNOW ABOUT ME	Quills are special, very sharp hairs with barbed tips. When they feel threatened, porcupines make their quills stand up as a warning that attackers will get a face full of needles. Porcupines move slowly and can't see well, so those quills protect them from predators. The quills of eastern porcupines are more black and brown. In the West, they are more yellow.

RACCOON

Procyon lotor

Gray, furry animal with a black face mask and ringed tail

WHERE DO I LIVE?

ID ME	SIZE Body about 1.5 feet (46 cm) long; tail another 10 to 14 inches (25 to 35 cm)
	SIGNS Tracks look like 3-inch (75 cm) hands; doglike scat full of seeds; ripped up holes in lawn
LOOK FOR ME	Woods, wetlands, along rivers and creeks, suburbs, and parks
WHAT I EAT	Crayfish, fruit, small prey, worms, grain, crops, and garbage
SOME THINGS TO KNOW ABOUT ME	Raccoons are adaptable omnivores. That means they eat whatever they can find!
	Their front paws have nimble fingers that are able to open trash cans and coolers and feel along the bottom of a stream.
	Raccoons are nocturnal and usually sleep away the day in a tree.

MAMMALS

RED FOX

....

Vulpes vulpes

....

Rusty-red fox with blackish legs and a bushy tail with a white tip

WHERE DO I LIVE?

ID ME	**SIZE** Body about 2 feet (61 cm) long; tail another 12 to 17 inches (30 to 43 cm)
	SIGNS Tracks in a straight line, like a cat, but with claw points; dark and pointy droppings with fur, seeds, and bug parts
LOOK FOR ME	Pastures and fields, woodlands, parks, and suburbs
WHAT I EAT	Rabbits and voles in winter; fruit and insects in summer; also birds, mice, and earthworms
SOME THINGS TO KNOW ABOUT ME	Red foxes look like small dogs but move more like cats. They ambush their prey and kill with a jumping pounce.
	Male-female pairs live together in a territory and raise families together. Each litter has between four and seven young, called kits.
	Red foxes can run 30 miles (48 km) per hour and jump over things that are 6.5 feet (2 m) tall. What athletes!

STRIPED SKUNK

Mephitis mephitis

Furry, black-and-white animal the size of a cat with short legs and a wide, bushy tail

WHERE DO I LIVE?

ID ME	**SIZE** Body 13 to 16 inches (30 to 40 cm) long; tail nearly as long
	SIGNS Stinky smell, 1.5-inch (4 cm) tracks with five toes; ripped up holes in lawn
LOOK FOR ME	Woods, grasslands, and suburbs
WHAT I EAT	Insects, bird eggs, carrion, fruit, and some small mammals
SOME THINGS TO KNOW ABOUT ME	Skunks can hit an intruder with their smelly, oily spray from 10 feet (3 m) away. They shoot it out of two glands under the tail. A raised tail gives warning, so watch out!
	There a several species of skunks with different coat patterns. Striped skunks are black with two thick white stripes on the back that come together in a V at the neck.
	All skunks are nocturnal and spend days sleeping in rock piles, hollow logs, old woodchuck holes, or under sheds.

MAMMALS

THIRTEEN-LINED GROUND SQUIRREL

Ictidomys tridecemlineatus

Small-eared, tan ground squirrel with lines of stripes and spots along body, often standing on hind legs

WHERE DO I LIVE?

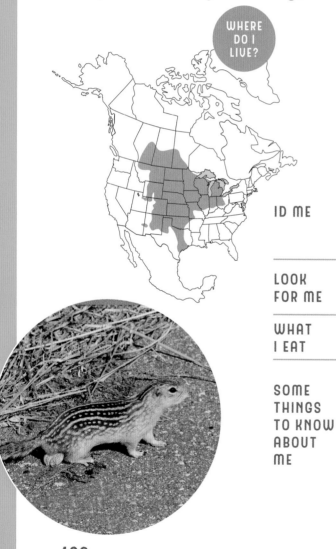

ID ME	**SIZE** Body about 5 inches (13 cm) long; tail another 3 inches (7.5 cm)
	SIGNS Small burrow hole with lots of pathways to it; trilled whistle sound
LOOK FOR ME	Mowed grassy areas like pastures, lawns, cemeteries, and golf courses
WHAT I EAT	Seeds, grasshoppers, and crickets
SOME THINGS TO KNOW ABOUT ME	Count the stripes — there really are 13! Some are solid and others are chains of white spots.
	Ground squirrels are diurnal (active during the day), making them easy to see.
	Winters are spent in deep hibernation with a body temperature lowered to near freezing. Their heart beats slow down to only 4 or 5 times per minute, instead of the usual 200 beats.

VIRGINIA OPOSSUM

Didelphis virginiana

Gray, furry animal with a long naked tail and white pointy face

WHERE DO I LIVE?

ID ME	**SIZE** Body 15 to 19 inches (38 to 48 cm) long; tail another foot (30 cm) or longer
	SIGNS Handlike tracks about 2 inches (5 cm) in size; bright, yellow-green shining eyes at night
LOOK FOR ME	Forests, farmlands, and suburbs
WHAT I EAT	Fruit, grubs, carrion, and garbage
SOME THINGS TO KNOW ABOUT ME	The Virginia opossum is the only marsupial that lives north of Mexico. Marsupials are mammals (like kangaroos) whose babies develop inside a pouch while each is attached to a nipple.
	Where winters are cold, opossums often lose their tail tips and ears to frostbite.
	Opossums don't move very fast, but they are great tree climbers. Their strong prehensile (grasping) tails can hold on to branches.

MAMMALS

WHITE-TAILED DEER

Odocoileus virginianus

Large, brown deer flashing a white tail as it runs

WHERE DO I LIVE?

ID ME	**BODY** About 6 feet (2 m) long **SIGNS** Tracks from narrow split hooves; hard, dark scat pellets about 1 inch (2 cm) long; small trees with missing bark from males rubbing antlers; patches of flattened grass or snow from napping deer; clipped flowers and garden plants; shed antlers
LOOK FOR ME	Suburbs, woods, farmland, brushy areas, and forest edges
WHAT I EAT	Plants, leaves, twigs, shoots, acorns, berries, seeds, and some grasses
SOME THINGS TO KNOW ABOUT ME	By 1900 most of the eastern United States was deer-less because of overhunting. But after a century of controlled hunting and land conservation (and few remaining predators), there are more white-tailed deer in the United States today than before European settlers arrived. Only male white-tailed deer have antlers. They shed them every year in late winter to early spring and then grow new ones. Baby white-tailed deer, or fawns, are born with white spots that help camouflage them. For the first month or so, the mother leaves the fawn in hiding while she finds food, coming back only to nurse her baby.

WOODCHUCK

....

Marmota monax

....

Big, husky, grizzled brown ground squirrel with little ears and a stubby tail

WHERE DO I LIVE?

ID ME	BODY 13 to 26 inches (33 to 66 cm) long
	SIGNS Large burrow with a 10-inch (25 cm) wide hole; loud single-note whistle
LOOK FOR ME	Meadows, edges of forests, pastures, and suburbs
WHAT I EAT	Grasses and plants
PLANT THIS FOR ME	Native clovers and sorrels
SOME THINGS TO KNOW ABOUT ME	Whether you call it a groundhog, a whistlepig, or a woodchuck, it's North America's largest member of the squirrel family.
	Everyone from skunks and raccoons to rabbits and foxes use old woodchuck burrows.
	Woodchucks hibernate in winter. Deep in a burrow and curled up in a ball, their heart slows to only a few beats a minute.

MAMMALS

WILDLIFE WORDS TO KNOW

AMPHIBIANS. Cold-blooded animals that live in water and breathe with gills when young, then develop lungs and live on land as adults; frogs, toads, salamanders, and newts.

ARACHNIDS. Small invertebrate animals that mostly live on land, have eight legs, no wings, and no antennae; spiders, scorpions, mites, ticks, and harvestmen.

BIODIVERSITY. When many different species live in an area.

BUGS. See *insects*.

CAMOUFLAGE. Colors and/or patterns that help something blend in with its surroundings, making it hard to see.

CARRION. The flesh of dead animals, a source of food for some other animals.

COLD-BLOODED. Having a body temperature that changes with the temperature of the surroundings.

COLONY. A group of animals of the same species that live together.

CONSERVATION. Protecting and maintaining natural places.

COVER. Places for animals to find shelter and hide from predators and/or prey.

CRUSTACEANS. Invertebrate animals with a shell, jointed legs, gills, and antennae; includes lobsters, crabs, shrimps, crayfish, pill bugs, and wood lice.

DIURNAL. Active during the day and sleeping at night.

ECOSYSTEM. A community of plants and animals interacting with each other and their environment.

HABITAT. The place where a plant or animal naturally lives and can survive.

HIBERNATE. To pass the winter in a sleeping or resting state.

HOST PLANT. A particular plant that an animal, like a butterfly caterpillar, needs to eat during part of its life cycle.

INSECTS. Small invertebrate animals with six legs, usually wings, and three body parts; flies, grasshoppers, bees, butterflies, and beetles.

INVASIVE SPECIES. A non-native plant or animal that crowds out the species that live in an area.

INVERTEBRATE. An animal without a backbone; worms, insects, spiders, and slugs.

LARVA (PLURAL LARVAE). A young animal with a different form from an adult; for example, tadpoles, caterpillars, and maggots.

MAMMALS. Animals that are warm-blooded, have a backbone and hair, and feed their young with milk from the mother's body.

MIGRATE. To relocate from one habitat to another in a regular cycle.

MOLLUSKS. Invertebrate animals with soft bodies sometimes covered in hard shells; snails, slugs, clams, mussels, octopuses, and squids.

NATIVE. Originally from a place, rather than coming from or being brought from somewhere else.

NOCTURNAL. Active at night.

NON-NATIVE. Not originally from a particular place.

PESTICIDES. Chemicals used to kill unwanted insects, plants, pest animals, or fungi.

POLLINATION. When flowers are fertilized by pollen carried by insects, wind, birds, or bats.

POLLUTION. Harmful substances in the air, water, and soil because of human activities.

PREDATORS. Animals that kill and eat other animals.

PREY. An animal that is hunted by another animal for food.

REPTILES. Cold-blooded animals — including snakes, turtles, lizards, and alligators — with no legs or short legs and a body covered in scales or bony plates.

SCAT. Animal droppings; also called feces or poop.

SCAVENGER. An animal that eats dead animals and plant material.

SPECIES. A group of plants or animals that share characteristics and are able to breed and reproduce their own kind.

VERTEBRATE. An animal with a backbone; fish, reptiles, amphibians, birds, and mammals.

WARM-BLOODED. Maintaining a constant body temperature; birds and mammals are warm-blooded.

FIND OUT MORE: RESOURCES

Got questions? Here are some websites (w), books (b), and apps (a) to check out if you want to know more about the birds, bugs, bats, and other creatures you've seen, or you need help with an ID or more info on native plants.

ALL ANIMALS

(w) Animal Diversity Web
University of Michigan Museum of Zoology
https://animaldiversity.org

(w) Encyclopedia of Life
Smithsonian National Museum of Natural History
http://eol.org

(a) iNaturalist
www.inaturalist.org

(w) (a) Map of Life
https://mol.org

(w) (a) MyNature
www.mynatureapps.com

(b) *National Geographic Illustrated Guide to Wildlife* by Catherine Herbert Howell (National Geographic Books)

BIRDS

(w) All About Birds Bird Guide
The Cornell Lab of Ornithology
www.allaboutbirds.org/guide

(a) BirdsEye
www.birdseyebirding.com

(w) (a) eBird
The Cornell Lab of Ornithology
https://ebird.org

(a) Merlin
The Cornell Lab of Ornithology
http://merlin.allaboutbirds.org

HEAR & LEARN BIRD SONGS

(w) All About Birds
The Cornell Lab of Ornithology
www.allaboutbirds.org/
how-to-learn-bird-songs-and-calls

(w) Bird-Sounds.net
www.bird-sounds.net

MAMMALS

(w) Bat Conservation International
www.batcon.org

(b) *Mammals of North America* by Roland W. Kays and Don E. Wilson, 2nd ed. (Princeton University Press)

BUGS

(b) *Backyard Guide to Insects & Spiders of North America* by Arthur V. Evans (National Geographic)

(w) BugGuide
Iowa State University
https://bugguide.net
An easy-to-use clickable guide and lots of photos help you identify bugs.

(w) Songs of Insects
http://songsofinsects.com
Listen to cricket, katydid, and cicada calls.

(w) Xerces Society for Invertebrate Conservation
https://xerces.org
Learn all about helping native bees, butterflies, and other pollinators.

REPTILES & AMPHIBIANS

(w) AmphibiaWeb
https://amphibiaweb.org

(w) The Reptile Database
www.reptile-database.org

HEAR & LEARN FROG AND TOAD CALLS

(w) Frog Call Quizzes
Patuxent Wildlife Research Center
US Geological Survey
www.pwrc.usgs.gov/frogquiz

(w) The Songs of Frogs and Toads
www.mister-toad.com/frogcalls.html

PLANTS & POLLINATORS

(w) (b) Bringing Nature Home
www.bringingnaturehome.net
The website of Doug Tallamy, professor and author of *Bringing Nature Home: How You Can Sustain Wildlife with Native Plants* (Timber Press).

(w) Native Plant Finder
National Wildlife Federation
www.nwf.org/NativePlantFinder
Enter your zip code to discover the best native plants for wildlife where you live.

W Lady Bird Johnson
Wildflower Center
www.wildflower.org/collections
Find native plant lists by state or province,
as well as plants that attract pollinators and
beneficial insects.

W Pollinator Conservation
Resource Center
The Xerces Society for Invertebrate
Conservation
https://xerces.org/pollinator-resource-center

a Leafsnap
http://leafsnap.com

Do More: Citizen Science

Want to help wildlife beyond your yard? Take it to the next level! Here are the web addresses of some great projects that put regular people to work helping map, track, count, and aid wild animals of all kinds. Citizen scientists rock!

W Audubon Christmas Bird Count
www.audubon.org/conservation/science
/christmas-bird-count
Join thousands of birders in two yearly
counts of our feathered friends.

W Bumble Bee Watch
https://www.bumblebeewatch.org
Upload photos of bumblebees to help find
out where they're in trouble.

W FrogWatch USA
www.aza.org/frogwatch
Join up with a local chapter to find out how
to help frogs and toads.

W The Great Backyard Bird Count
http://gbbc.birdcount.org

W Journey North
www.learner.org/jnorth
Track migrations of hummingbirds, American
robins, orioles, swallows, cranes, gray
whales, butterflies, and bald eagles, as well
as report on when trees leaf out and frogs
start singing. There's a project for everyone!

W Lost Ladybug Project
www.lostladybug.org/participate.php
Which kinds of ladybugs are thriving where
you live? Help find out by uploading your
photos.

W NestWatch
https://nestwatch.org
Experts need information on the health of
nesting birds across the continent. You can
help by monitoring bird nests in your area.

W Project FeederWatch
https://feederwatch.org
Identify and count birds coming to your
feeders during winter.

W Western Monarch
Milkweed Mapper
www.monarchmilkweedmapper.org
Submit sightings of monarch butterflies west
of the Rocky Mountains.

Download It

W Bat House
www.batcon.org/resources
/getting-involved/bat-houses
Find everything you need to know to build a
successful bat house from Bat Conservation
International.

W Bumblebee Box
www.xerces.org/fact-sheets
Click on Nests for Native Bees under Bee
Nests to download plans for a bumblebee
box, as well as other native bee nests from
the Xerces Society.

W Pollinator-Friendly Plant Lists
https://xerces.org/pollinator-conservation
/plant-lists
Choose the list for your region, download,
and get planting.

W Wild Notes Notebook
https://www.storey.com
/wildlife-ranger-downloads/
Download pages for your notebook.

W Window Cling Patterns
https://www.storey.com
/wildlife-ranger-downloads/
Download fun patterns for making window
clings.

W Wooden Bird Nest Box
The Cornell Lab of Ornithology
https://nestwatch.org/learn
/all-about-birdhouses/
Choose a bird and download plans to build
it a nest box. And there's info on adding a
webcam!

INDEX

italic = illustration of an animal or its tracks